Surface Mount Technology
Terms and Concepts

Surface Mount Technology Terms and Concepts

Phil Zarrow and Debra Kopp

Newnes
Boston Oxford Johannesburg Melbourne
New Delhi Singapore

Newnes is an imprint of Butterworth–Heinemann.

Copyright © 1997 by Butterworth–Heinemann
A member of the Reed Elsevier group
All rights reserved.

No part of this publication may be reproduced, stored in a retrieval system, or transmitted in any form or by any means, electronic, mechanical, photocopying, recording, or otherwise, without the prior written permission of the publisher.

Recognizing the importance of preserving what has been written, Butterworth–Heinemann prints its books on acid-free paper whenever possible.

Butterworth–Heinemann supports the efforts of American Forests and the Global ReLeaf program in its campaign for the betterment of trees, forests, and our environment.

Library of Congress Cataloging-in-Publication Data
Zarrow, Phil, 1953–
 Surface mount technology terms and concepts / Phil Zarrow and Debra Kopp.
 p. cm.
 ISBN 0-7506-9875-6 (alk. paper)
 1. Surface mount technology—Terminology. I. Kopp, Debra. II. Title.
 TK7870.15.Z374 1997
 621.3815'31—dc21   97-14824
                                             CIP

British Library Cataloguing-in-Publication Data
A catalogue record for this book is available from the British Library.

The publisher offers special discounts on bulk orders of this book.
For information, please contact:
Manager of Special Sales
Butterworth–Heinemann
313 Washington Street
Newton, MA 02158–1626
Tel: 617-928-2500
Fax: 617-928-2620

For information on all Newnes electronics publications available, contact our World Wide Web home page at: http://www.bh.com/newnes

10 9 8 7 6 5 4 3 2 1

Printed in the United States of America

Acknowledgments

We wish to thank all of our clients, students, and colleagues who encouraged us to produce this glossary of terminology. We hope it will help dispel some of the uncertainty and confusion regarding terms and abbreviations used in our industry.

A-stage The condition of low molecular weight of a resin polymer during which the resin is readily soluble and fusible.

absorbtivity The percentage of infrared absorbed by a substrate as compared with the total incident infrared.

acceptance tests A set of tests performed to determine the acceptability of a board or assembly.

access holes A series of holes in successive layers, each set having a common center or axis. These holes in a PCB provide access to the surface of the land in one of the layers of the board.

accuracy (1) The difference between measured results and target values. (2) The difference between the measured result and the target value.

activated Condition in which a mixture has a higher chemical activity than normally found.

activated flux Rosin- or resin-based flux, with one or more activators added.

activating A treatment that renders nonconductive material receptive to electroless deposition.

activator Additive in a flux that aids the flux's cleaning ability.

active component Electronic components such as transistors and diodes that can operate on an applied signal, changing its basic character.

active temperature The ratio of the actual temperature to the melting temperature of solder.

adaptor A device that locates and supports products to be tested.

additive process A process for obtaining conductive patterns by the selective deposition of conductive material on unclad base metal.

adhesion The force of attraction between unlike molecules.

adhesive A material used for bonding, sealing, and joining laminates, films, foils, coils, and conductors.

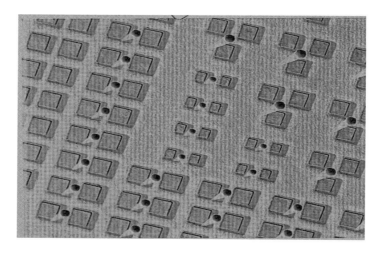

Figure 1 Adhesive deposited for component attachment. The attached components will then be wave soldered. (Courtesy of Camelot Systems, Inc.)

aerosol   Liquid particles small enough to be airborne.

air leveling   A process used in solder dipping of bare copper circuitry in which high-velocity air is used to blow solder clear from plated-through holes and minimize solder thickness. Also known as hot air leveling (HAL) and hot air solder leveling (HASL).

alloy   A combination of two or more elements whereby at least one is a metal.

ambient   The surrounding environment that contacts the system, assembly, or component of interest.

ambient temperature   Dry bulb temperature of the surrounding atmosphere.

amorphous   Condition where atoms and molecules of a material are not arranged in a definite pattern or form (non-crystalline). Amorphous materials typically lack certain well-defined physical properties such as a distinct melting point or boiling point. These materials usually have poor thermal and electrical conductivity properties. Examples include glass, carbon, and rosin.

analog circuits   Circuits that provide a continuous relationship between input and output.

analog functional testing   A functional testing of a PCB at board level. Various analog test signals are applied to the board through a switch or multiplexor. Most effectively used on analog and hybrid boards.

analog in-circuit testing   A system of testing that measures component values on a loaded PCB before power is applied. Best used for analog or hybrid boards.

analog test   A functional test of a PCB at board level in which signals are applied through a switch to pinpoint outputs.

angle of attack The angle between the squeegee face and the plane of the stencil.

anion A negatively charged atom or radical.

anisotropic adhesive A material filled with a low concentration of large conductive particles designed to conduct electricity in the Z-axis but not the X- or Y-axis. Also called a Z-axis adhesive.

annotation Text or legend pertinent to a board design. Text appears off the board area and consists of lettering and symbols, while legend appears on the board.

annular ring The portion of conductive material completely surrounding a hole.

anode Positive pole of a plating cell from which negatively charged ions leave the plating solution by conversion back to their parent atoms or atom groups.

ANSI American National Standards Institute

anti-static material An electrostatic discharge (ESD) protective material having a surface resistivity greater than 10^9 but not greater than 10^{14} ohms per square centimeter.

aqueous cleaning A water-based cleaning methodology that may include the addition of the following chemicals: neutralizers, saponifiers, surfactants, dispersants, and anti-foaming agents.

array A group of elements arranged in rows and columns.

artwork The image of the PCB conductive pattern that then generates the artwork master. It is generally scaled at either 3:1 or 4:1, but can be made at any scale. Accuracy in scaling is imperative because PCBs will be less accurate than artwork by virtue of the scale factor.

artwork generation The process of transferring the idea for a circuit pattern into a precise, reproducible artwork master for mass production manufacturing. Generation can be executed by the traditional manual drafting and photographic technique or via electronic means.

artwork master The photographic film or plate that embodies the image of the PCB pattern, usually on a 1:1 scale. Commonly a sensitized gelatin plate, but for high precision work, a chromium plate can be used.

artwork registration system Equipment of various sizes and complexities used to achieve artwork registration. Accuracy, repeatability of precision tolerances, loading and unloading simplicity, and speeds are significant aspects.

aspect ratio A ratio of the PCB thickness to the diameter of the smallest hole.

assembly Refers to inserting, placing, and joining the components or subassemblies to a bare board surface.

assembly drawings Documents that depict the physical relationship of two or more parts, their combination and accompanying assemblies.

ASTM American Society for Testing and Materials

atom Smallest particle of an element that can enter into a chemical combination.

auto-ignition point Temperature where vapor from a material in air will burst spontaneously into flame.

automated optical inspection (AOI) Test fixture method in which printed circuit boards are checked at bareboard, pre- or post-soldered stages of assembly by optical means.

automatic component insertion Insertion of components into a through-hole printed circuit board by automatic means.

automatic debridging A process in which a hot air knife removes excess solder and exposes poor solderability conditions. Used largely in stress testing and wave soldering.

automatic test equipment (ATE) Equipment designed to automatically analyze functional or static parameters in order to evaluate performance degradation. It may also be designed to perform fault isolation.

automatic test generation (ATG) Computer generation of a test program based solely on the circuit topology, requiring little or no manual programming effort.

available hours The actual number of hours available in a shift after taking into consideration breaks, meetings, etc.

axial leads Leads coming out of the ends of discrete component or device along the central axis.

azeotrope A mixture of two or more polar and non-polar solvents that behaves like a single solvent to remove polar and non-polar contaminants. The boiling point is lower than its individual components and reaches one boiling point temperature as if it were a single component solvent.

B-stage The condition of resin polymer when it is more viscous, with higher molecular weight, and insoluble, but plastic and fusible.

B-stage material Sheet material impregnated with a resin cured to an intermediate stage. Also referred to as prepreg.

back bonding Bonding active chips to the substrate using the back of the chip, leaving the face with its circuitry face up. The opposite process is face down bonding.

backplanes and panels Interconnection panels into or onto which printed circuits, other panels, or integrated circuit packages can be plugged or mounted.

ball grid array (BGA) Integrated circuit package in which the input and output points are solder bumps arranged in a grid pattern. Also known as bumped grid array.

bare board A printed circuit board that has all lines, pads, and layers completed but that has no components installed.

bare board testing Testing procedure in which printed circuit boards are tested prior to assembly of components.

base (1) A substance dissolved in water that produces hydroxyl ions comprised of an oxygen atom and a hydrogen atom. (2) Insulating layer of a printed circuit board.

base copper The thin copper foil portion of a copper-clad laminate for PCBs that can be present on one or both sides of the board.

base material The material (rigid, rigid-flex, or flexible) that serves as the substrate or foundation of the conductive pattern of the PCB.

bed-of-nails technique A method of testing production volume PCBs based on the theory that most failures that are the result of manufacturing defects (e.g., solder bridges, reversed components, etc.) or defective components can be caught and identified during in-circuit component testing. A test fixture, heavily laden with nail-type contact probes, contacts points on the board allowing the devices to be powered up. Then, using active guarding techniques, the system quickly determines the faulty devices or connections.

bed-of-nails test fixture A vacuum-activated or mechanically activated test fixture comprised of a frame, holder, and a field of spring-loaded probes designed to make electrical contact with the assigned PCB nodes.

BGA Abbreviation for ball grid array or bumped grid array.

blackbody A body that absorbs all radiation incident upon its surface.

blade attack angle The angle formed between the stencil and squeegee blade during squeegee travel.

bleeding A condition in which a plated hole discharges process material or solution from crevices or voids.

blind via A via that connects an outer layer of a printed circuit board to an inner layer but does not continue through to the other side. Also known as a buried via.

blister Delamination that occurs in localized areas of the PCB.

blistering The lifting of a copper track from a printed circuit board surface after soldering.

blow hole A void caused by outgassing.

board The unpopulated printed circuit board. The foil and base from which a printed circuit is fabricated.

board thickness The thickness of the metal-clad base material, including conductive layer or layers. May include additional platings and coatings depending on when the measurement is made.

bond liftoff The failure mode whereby the bonded lead separates from the surface to which it was bonded.

bond strength Strength of adhesion between two joined materials. Also known as peel strength.

bonding agent The adhesive used to bond individual layers into a laminate.

bonding layer An adhesive layer used in bonding together other discrete layers of a multi-layer PCB during lamination.

bonding pad A metallized area at the end of a thin metallic strip to which a connection is to be made. Also called a bonding island.

bonding time The time from the commencement of hot bar heat-up (moment of hot bar contact) until the reflow profile is completed.

bonding wire Fine gold or aluminum wire between bonding pads on a semiconductor and base lands.

bootstrap A feedback technique that tends to improve linearity and input impedance of circuits operating over a wide range of input signals.

boss On a printed circuit, it is the conductive area to which components or separate circuits are attached. A plated-through hole land is the conductor surrounding the hole through the conductive pattern and the base material. Also called land, pad, terminal point, tab, spot, or donut.

boundary scan A self test designed into components at the silicon level that allows testing by means of a built-in four or five pin test bus accessing input and output pins.

bow The deviation from flatness of a board characterized by a roughly cylindrical or spherical curvature such that if the board is rectangular, its four corners are in the same plane.

branched conductor Any conductor on a PCB that connects electrically to more than two leads.

breadboard model An assembly designed to prove the feasibility of a circuit, system, or principle.

breakaway panels Hardboards held together in a grouping with breakaway tabs. The panels make handling and processing easier for automatic placement and soldering; the boards can be separated by snapping apart.

bridge Solder that "bridges" across two conductors that should not be electrically connected, i.e., causing an electrical short.

brush fluxing A specialized wave solder technique in which a 360° bristled brush rotates in a foaming flux head to transfer the flux to the board.

bumped grid array Integrated circuit package in which the input and output points are solder bumps arranged in a grid pattern. Also known as ball grid array and abbreviated BGA.

bumps Inner terminations of a tape automated bonded and flip-chip integrated circuit.

buried via A via that connects two inner layers of a printed circuit board and hence is not visible from the outer layers. Also known as a blind via.

burn-in The operation of items prior to their ultimate application. Intended to stabilize their characteristics and to identify early failures. The process is designed to accelerate the aging of a device.

burn-in testing A method used to detect mortality in devices prior to their being put in subassemblies or into actual service. Usually occurs before the devices are potted or connected to a permanent heat sink.

bus A circuit over which data or power is transmitted from any of several sources to any of several destinations between circuits.

bus bars Power distribution components, many of which consist of two or more conductive layers electrically insulated from one another and from other components by thin dielectric layers.

butt joint A solder joint formed with a "footless" lead also known as an I-lead.

C

C-stage The condition of a resin polymer when it is in the solid state, with high molecular weight, being insoluble and infusible.

CAD/CAM system Computer-aided design refers to the use of computers in formulating designs. Computer-aided manufacturing refers to the use of computers to translate the design into an actual product. CAD/CAM systems are usually modular, with hardware and software programs selected to user requirements. Typically, these systems include a computer and mass memory for information processing and storing, input terminals for design creation and observation, and output devices for converting stored information into drawings and reports. CAD/CAM systems are used in the electronics industry to perform all steps of artwork generation (from concept to actual production), specifically in the manufacturing of circuits.

camber The planar deflection of a flat cable or flexible laminate from a straight line or specified length. A flat cable or flexible laminate with camber is similar to the curve of an unbaked race track.

capacitance That property of a system of conductors and dielectrics that permits the storage of electricity when potential difference exists between the conductors.

capillary action The combination of force, adhesion, and cohesion that causes liquids such as a molten metal to flow between closely spaced solid surfaces against the force of gravity.

card guide A plastic or metal support for PCBs. It relieves the stress on connector contacts, makes insertion into and extraction from the connector easier, and eliminates the possibility of twisting the board.

carriers Holders for electric parts and devices that facilitate handling during processing, production, imprinting, or testing operations.

castellation A leadless termination used on LCCC packages for connection to the PCB.

catalyst A substance which accelerates the cure of a resin without being altered in the reaction.

Celsius A temperature reference related to Fahrenheit by $°C = 5/9 \times (°F - 32)$. Formally called centigrade.

center board support A mechanism employed in reflow ovens to prevent excessive PCB warpage during the reflow process. (See Figure 2 on following page.)

center-to-center spacing The nominal distance between the centers of adjacent features on any single layer of a PCB.

central component orientation A design concept in which the predominant components are centered in an area with the supporting components radiating outward. This method is applicable when there are one or more complex multi-lead components.

ceramic substrates These materials are used primarily because of their low loss qualities, long life characteristics, and ability to withstand high operating temperatures and heat shock. For general application, the aluminum oxide substrate (alumina) is preferred because of its availability, low cost, good thermal expansion coefficient to most inks. Beryllia substrates with a high thermal conductivity may be used in high power circuitry.

14 certification

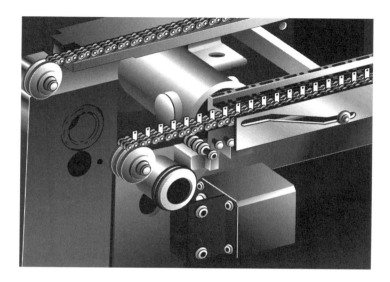

Figure 2 Center board support as used on reflow soldering oven to minimize PCB warpage. (Courtesy of BTU International)

certification Verification that specified testing has been performed and required parameter values have been attained.

CFC Chlorinated fluorocarbon.

chamfer Angle on the leading edge of a printed wiring board that permits easier insertion of the board into the connector, or angle on the inside edge of the barrel entrance of a connector which permits easier insertion of the cable.

characteristic impedance The ratio of voltage to current in a propagating wave, i.e., the impedance which is offered to the wave at any point of the line. In printed wiring its value depends on the width of the conductor to ground plane(s), and the dielectric constant to the media between them.

checklist The specified criteria compiled for evaluation during an inspection.

chemical hole cleaning The chemical process for cleaning conductive surfaces exposed within a hole. See also etchback.

chip A single substrate on which all the active and passive circuit elements have been fabricated using one or all of the semiconductor techniques of diffusion, passivation, masking, photoresist, and epitaxial growth. A chip is not ready for use until packaged and provided with external connectors.

chip capacitors Discrete devices that introduce capacitance into an electronic circuit, made in tiny wedge or rectangular shapes.

chip carrier A high density integrated circuit packaging technique in which input and output terminals are around the perimeter of the device, instead of only on two sides, as in dual in-line packages.

chip component Refers to leadless ceramic or plastic discrete capacitors, resistors, and inductors.

chip resistors Chip resistors are small chips of ceramic. The ceramic chip is an inert substrate with the resistor on the surface. Their small size is their chief virtue but by no means their only virtue. Chip resistors have extremely low shunt or parasitic capacitance, no inductance, are stable, and generally low in cost.

chip scale package Active, multi-I/O package that is no larger than 125% of the size of the silicon integrated circuit chip.

16 chip scale package

Figure 3 Encapsulated chip-on-board (COB). (Courtesy of Camelot Systems, Inc.)

chip testers Usually large, complex computer-based testers that test individual ICs, especially LSI and VLSI.

chip-and-wire A hybrid technology employing face-up-bonded chip devices exclusively interconnected to the substrate conventionally, i.e., by flying wires.

chip-on-board (COB) A hybrid technology exclusively employing face-up-bonded chip devices interconnected to the substrate conventionally, i.e., by flying wires. A generic term for mounting an unpackaged silicon die directly onto the PCB. Connections can be made by wire bonding, tape automated bonding (TAB), or flip-chip bonding.

chlorinated hydrocarbon An organic compound having hydrogen and chlorine atoms in its structure, such as methyl chloroform (1,1,1 trichloroethane) and used as a cleaner.

circuit The interconnection of a number of devices in one or more closed paths to perform a desired electrical or electronic function.

circuit card assembly (CCA) Term for a PWB after all electrical components have been attached. Also see printed circuit assembly and printed wiring assembly.

circuit density Number of circuits on a given area of a printed circuit board.

circuit mil area (CMA) A unit of area equal to the area of a circle whose diameter is one mil (0.001"). Used chiefly in specifying cross-sectional areas of conductors.

circuit tester A method to properly test production volume PCBs, such as by bed-of-nails, footprint, guided probe, internal trace, loaded board, bare board, or component testing.

18 circuit verifier

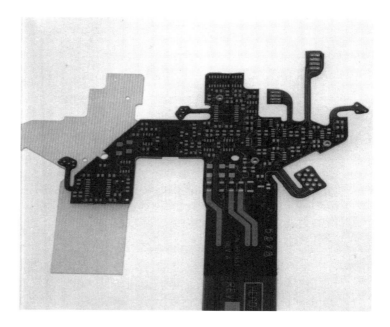

Figure 4 Flexible circuit board. (Courtesy EDN Magazine)

circuit verifier An assembly that electrically stimulates an item being tested and monitors proper response. This includes shorts and opens testing, as well as functional diagnostic testing. Also called analyzer, test set, and tester.

clad or cladding A relatively thin layer or sheet of metal foil bonded to a laminate core to form the base material for printed circuits.

clamshell fixture An in-circuit test fixture designed to probe both sides of a PCB. The top probe plate is hinged to alloy the PCB to be inserted. Also referred to as a clamshell tester.

clean room A special manufacturing area where the air is filtered to remove dust particles and precautionary measures are used to keep contamination away from the unprotected circuit during processing.

clearance The space between the punch and walls of the die plate. This space has a direct relation to part or hole sizes.

clearance hole A hole in the conductive pattern larger than, but coaxial with, a hole in the PCB base material.

clinched leads Component leads that extend through the PCB and are formed to effect a spring action, metal-to-metal electrical contact with the conductive pattern prior to soldering.

clinching Process of crimping through-hole component leads underneath the printed circuit board to maintain the component in place.

close center probing Test points on a product that are located on centers closer than 0.050".

cluster test An in-circuit test method in which a group of components are checked as a unit.

coatings Material used for a wide variety of purposes, such as moisture protection, bonding of windings into integral masses, improvement of electrical masses, improvement of electrical properties, protection against chemicals and mechanical abuse, appearance, etc. Some printed circuit coatings facilitate repairability by selectively stripping the coating with chemical solvents. Examples include various resins, varnishes, epoxies, lacquers, and phenolic materials.

coefficient of thermal expansion The fractional change in dimension of a material for a unit change in temperature. Expressed in parts per million per °C. Also known as CTE.

cohesion The force of attraction between like molecules.

cold cleaning An organic solvent cleaning process in which liquid contact accomplishes the solution and removal of residues after soldering.

cold short A brittle condition in metal at temperatures that are below the recrystallization temperature.

cold solder joint A solder connection exhibiting poor wetting and a grayish, porous appearance due to insufficient heat, inadequate cleaning prior to soldering, or excessive impurities in the solder solution.

color selectivity Preferential absorption of radiation caused by emitted energy in the visible band wavelength (.39–.79 microns).

combinational test Test procedure using both in-circuit and functional test methods.

component A part or combination of parts mounted together to perform a design function(s).

component density The quantity of components placed on a PCB per unit area.

component hole A hole used for the attachment and electrical connection of component terminations, including pins and wires, to the PCB.

component lead The solid or stranded wire or formed conductor that extends from a component and serves as a mechanical and/or electrical connection that is readily formable to a desired configuration.

component side The side of the PCB on which most of the components or the active components are mounted. See also primary side.

composite board A completely laminated, multi-layer PCB.

computer-guided probe A fault-isolation technique based solely on good circuit data. The probe algorithm acts as the master instruction to an operator to probe various IC pins on the unit under test until it derives the final diagnosis and diagnostics.

condensation inert curing   Adhesive curing method using condensation inert heating as the heat transfer medium.

condensation inert heating   A general term referring to condensation heating where the part to be heated is submerged into a hot, relatively oxygen-free vapor. The part, being cooler than the vapor, causes the vapor to condense on the part transferring its latent heat of vaporization to the part. Also known as vapor phase.

condensation inert soldering   Reflow soldering method using condensation inert heating as the heat transfer medium.

conduction   Heat transfer that occurs within or between solids due to temperature gradients across the solids.

conductive adhesive   An adhesive material that has metal powder added to increase electrical conductivity.

conductive epoxy   An epoxy material (polymer resin) that has been made conductive by the addition of a metal powder, usually gold or silver.

conductive foil   A thin sheet of metal that may cover one or both sides of the base material and intended to form a conductive pattern.

conductive ink   In hybrid technology, the conductive paste used on thick film materials to form the printed conductor pattern. Usually contains metal, metal oxide, glass frit, and solvent.

conductive pattern   The configuration or design of the conductive material on the base material. Includes conductors, lands, and through connections when these connections are an integral part of the manufacturing process, such as additive.

conductor A thin conductive area on a PCB or internal layer, usually composed of path and lands to which component leads are connected. Also called a trace.

conductor base spacing The conductor spacing at the plane of the surface of the base material.

conductor base width The conductor width at the plane of the surface of the base material.

conductor layer The total conductive pattern formed upon one side of a single layer of base material. See also physical layer.

conductor spacing The distance between adjacent edges (not centerline to centerline) of isolated conductive patterns in a conductor layer.

conductor thickness The thickness of the conductor including all metallic coating.

conductor to hole spacing The distance between the edge of a conductor and the edge of a supported or unsupported hole.

conductor width The observable width of the pertinent conductor at any point chosen at random on the PCB, normally viewed vertically above unless otherwise specified.

conformal coating A thin protective coating applied to a PCB assembly that conforms to the configuration of the assembly.

connector A device providing electrical connections/disconnections that consists of a mating plug and receptacle that interconnects PCBs with cables, racks, or chassis. Commonly used PC connectors are edge, two-piece, and hermaphroditic.

contact spacing 23

Figure 5 PCB assembly connectors. (Courtesy EDN Magazine)

constraining core substrate A composite PCB consisting of epoxy-glass outer layers bound to a low thermal expansion core material. The core artificially constrains the expansion of the outer layers to match the expansion coefficient of ceramic chip carriers.

contact angle The angle of wetting between the solder fillet and the termination or land pattern. It is measured by constructing a line tangent to the solder fillet that passes through a point of origin located at the plane of intersection between the solder fillet and termination or land pattern. Also known as wetting angle.

contact spacing The distance between the centerlines of adjacent contact areas.

contamination Any foreign material such as dirt, oil, and epoxy glass on an SMC lead or pad or on a footprint pad that interferes with solderability.

continuity A continuous path for the flow of current in an electrical circuit.

continuity test A test procedure in which voltage is applied at two interconnected points to determine the presence or absence of current flow.

continuity testing A testing procedure wherein voltage is applied to two interconnected lands in order to observe the presence or absence of current flow. This process is repeated until all interconnections on the board have been tested.

controlled collapse chip connection (C4) A solder joint connecting a substrate and a flip-chip, where the surface tension forces of the molten solder supports the weight of the chip and controls the height of the joint.

convection Heat transfer that occurs at the interface of a solid and fluid or gas due to temperature differences.

convection, controlled Convection heat transfer in which the convective characteristics (such as flow, rate, velocity and temperature) are precisely controlled.

convection, enhanced Convection heat transfer that occurs by drawing or drafting the fluid or gas over the solid media.

convection, forced Convection heat transfer that occurs by forcing the fluid or gas over the solid media. Also known as convection dominant.

convection, free Convection heat transfer that occurs due to movement of air caused by density gradients near the solid surface.

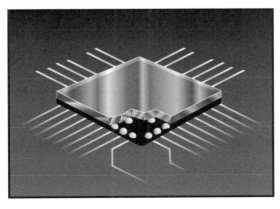

Figure 6 Direct chip attachment (DCA) using C4 (controlled collapse chip connection) methodology. Note underfill process. (Courtesy of Camelot Systems, Inc.)

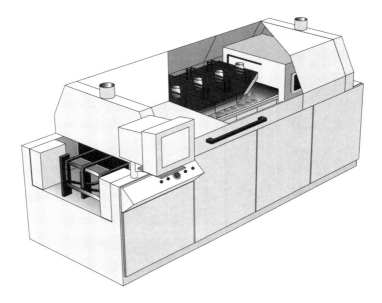

Figure 7 Forced convection (convection dominant) reflow soldering oven. (Courtesy BTU International)

convention A definite formatting method used in electronic diagrams to present the clearest picture of the circuit function. Some common conventions are (1) circuit signal flow from left to right with inputs on the left and outputs on the right; (2) various function stages of the circuit in the same sequence as the signal flow; (3) voltage potentials with the highest voltage at the top of the sheet and the lowest at the bottom; and (4) auxiliary circuits that are included, but are not a main part of the signal flow, such as oscillators and power supplies, on the lower half of the drawings.

conveyor, edge Conveying mechanism that supports the product by the edges.

conveyor, mesh Conveying mechanism that fully supports the product.

conveyor, secondary Conveying mechanism used beneath the edge conveyor. Used to catch fallen product in the reflow process.

cooldown The period in the reflow process, after peak temperature, during which the temperature falls to the point where the solder joints solidify or freeze.

coordinate tolerancing A method of tolerancing hole locations in which the tolerance is applied to linear and angular dimensions, usually forming a rectangular area of allowable variation.

coplanarity Measurement of the distance between the lowest and highest pin of a package when rested on a flat surface.

coplanarity error The maximum deviation of component leads away from perfect planarity. Determined by placing the package on a flat surface and measuring the worst-case lead height off the surface.

coplanarity test A measure of the distance between the lowest and highest pins of a device when rested on flat plane.

copper foil A cathode-quality electrolytic copper deposited as a thin, continuous sheet on rotating drums direct from refinery electrolytes. Used as a conductor for printed circuits, copper foil readily bonds to insulating substrates, accepts the printed resists, and etches out to make printed circuits. It is made in a number of weights (thicknesses). The traditional weights are 1 to 3 ounces per square foot (0.0014" to 0.0042" thick).

copper mirror test Test for corrosivity of flux to a thin copper film vacuum-deposited on a glass plate.

copper-clad invar A metallurgically bonded material, consisting of 36% nickel and 64% iron alloy, that is used in the fabrication of PCBs. It is normally employed to solve thermal expansion mismatch problems with surface-mounted devices and multi-layer boards.

corner mark The mark at the corners of the PCB artwork, the inside edges of which usually locate the borders and establish the contour of the board. Also called crop mark.

corrosion Slow destruction of materials by chemical agents and electrochemical reactions.

corrosive fluxes Fluxes consisting of inorganic acids and salts, generally required when the condition of the surface is not well-suited for rapid wetting by molten solder. Also called acid fluxes.

cosmetic defect A variation from the conventional appearance of an item, such as a slight change in its usual color.

cover layer, cover coat Outer layer(s) of insulating material applied over the conductive pattern on the surface of the PCB. Sometimes referred to as a "pads only" layer when a copper foil clad laminate is used.

cratering Wave soldering defect caused by minute quantities of gas originating in the base laminate and escaping through pin holes in the plating, thus blowing some of the molten solder out of the hole. This gives the board a faulty appearance, but an adequate joint usually remains.

crazing A condition that occurs internal to the laminated base material in which the glass fibers separate from the resin at the weave intersections. Crazing is visible as white spots or crosses below the base material surface and is usually a result of mechanically induced stress.

creep strength Resistance of a material to stretching and deformation.

crop mark The mark at the corners of the PCB artwork, the inside edges of which usually locate the borders and establish the contour of the board. Also called corner mark.

cross-sectional area of a conductor The sum of the cross-sectional areas of its component wires, that of each wire being measured perpendicular to its individual axis.

crosshatching The breaking of large conductive areas with a pattern of voids in the conductive material.

CTE Coefficient of thermal expansion.

cure To change the physical properties of a material by chemical reaction, by the action of heat and catalysts, alone or in combination, with or without pressure.

curing agent A chemical added to a thermosetting resin to stimulate curing. Also called hardener.

curing temperature The temperature at which a material is subjected to curing.

curing time In the molding of the thermosetting plastics, the amount of time it takes for the material to be properly cured.

current-carrying capacity The maximum current that can be carried continuously, under specified conditions, by a conductor without causing objectionable degradation of electrical or mechanical properties of the PCB.

cycle rate (1) A component placement term similar to the placement rate, except it measures the machine speed as it moves from the component pickup location to the board site location and back again without actually picking and placing components (a dry run). Also referred to as the test rate.

data file A database organized in a specific way for a specific application.

datum reference A defined point, line, or plane used to locate the pattern or layer for manufacturing, inspection, or for both purposes.

daughter board A PCB assembly that plugs into another PCB assembly or motherboard via a connector or lead terminations.

defect Any non-conformance with the normally accepted characteristics for a unit.

definition Accuracy of deposition of solder paste, solder mask and circuit track lines.

delamination A separation between any of the layers of the base material or between the laminate and the conductive foil, or both.

dendrite The individual metallic filaments resulting from electromigration.

dendritic growth A growth in the presence of condensed moisture and electrical bias resulting in metallic filaments between conductors. Also called whiskers.

deposit height The average height or depth of the deposited material.

deposition The process of applying a material to a base by means of vacuum, electrical, chemical, screening, stenciling, or vapor methods.

design automation The use of computers in the design process that are responsible for decision making and data manipulation.

design for manufacturability Designing a product to be produced in the most efficient manner possible in terms of time, money, and resources, taking into consideration how the product will be processed and utilizing the existing skill base (and avoiding the learning curve) to achieve the highest yields possible.

design width of conductor The width of a conductor as delineated or noted on the master drawing.

desoldering The process of disassembling soldered parts in order to repair, replace, inspect, or salvage them. Typical desoldering methods are wicking, pulse vacuum (solder sucker), heat and pull, and solder extraction.

device An individual electrical element, usually in an independent body, which cannot be reduced or altered without destroying its stated function.

device under test (DUT) Component, printed circuit board or assembly subjected to a test. Also known as unit under test (UUT) and loaded board.

dew point The temperature at which water vapors in the air begin to condense.

dewetting A condition that results when molten solder has coated a surface and then receded, leaving irregularly shaped mounds of solder separated by areas covered with a thin solder film and base metal is not exposed.

dielectric Any insulating medium that intervenes between two conductors. A material having the property that energy required to establish an electric field is recoverable in whole or in part as electric energy.

dielectric constant A measure of a material's ability to store electrical energy.

digital circuits Applied normally for switching operations where the output of the circuit assumes one or two states: a binary operation. However, three states are possible.

digitization The conversion of feature locations of a PCB design to its digital representation in X-Y coordinates.

dual in-line (DIL) Component shape with two parallel rows of connection leads.

dimensional stability Dimensional change caused by factors such as temperature, humidity, chemical treatment, age or stress, usually expressed as Δ units/unit.

dual in-line package (DIP) An integrated circuit package that has a row of pins for through-hole mounting to maintain information storage.

dip soldering A process whereby PCBs are brought in contact with the surface of a static pool of molten solder for the purpose of soldering the entire exposed conductive pattern in one operation.

direct chip attach Process by which the silicon die is mounted and interconnected directly to the PCB assembly as opposed to using a sub-carrier. This includes chip-on-board (COB) and flip-chip.

direct emulsion screen A screen whose emulsion is applied by painting directly onto the screen, as opposed to indirect emulsion type.

discrete components Individual components such as resistors, capacitors, transistors, and diodes.

dispersant A chemical added to water to improve its ability to remove particulate matter.

disturbed solder joint A soldering defect that results from motion between the joined members during solidification.

documentation Information accompanying a PCB that explains the basic electromechanical design concept, the type and quantity of parts and materials required, special manufacturing instructions, and up-to-date revisions. Three classifications of printed wiring documentation are: "minimum-used" for prototype and small quantity runs; "formal-used" for standard product line and/or production quantities; and "military-complies" with government contracts specifying procurement drawings for the manufacture of identical items by other than the original manufacturer.

donut On a printed circuit, it is the conductive area to which components or separate circuits are attached. A plated-through hole land is the conductor surrounding the hole through the conductive pattern and the base material. Also called boss, pad, terminal point, tab, spot, or land.

double-sided assembly A PCB with components mounted on both sides.

double-sided board A printed board with a conductive pattern on both sides.

down force The pressure exerted on the stencil during the print cycle, typically in the range of one to six pounds per squeegee inch. Also known as squeegee pressure.

download The ability of an analyzer to provide failure analysis and data logging to a host computer.

downtime The period of time a piece of equipment is not producing product due to scheduled (routine maintenance) or unscheduled (machine failure) events.

drag soldering Mass soldering process in which assembled boards are dragged across the surface of a bath of molten solder.

dragout When a printed circuit board, passing through a soldering or cleaning process, takes with it material used in the process (i.e., solder, solvent).

drawbridging A soldering defect in which a chip component has pulled into a vertical or near vertical position with only one termination soldered to the PCB. It is typically caused by force imbalances during the reflow soldering process. Also referred to as the Manhattan effect, tombstoning, and the stonehenge effect.

dross Metal oxides and other entrapped impurities that float in or on the surface of a molten metal bath. In the case of solder, it would include the oxides of lead and tin.

dry film resists Coating materials specifically designed for use in the manufacture of PCBs and chemically machined parts. They are suitable for all photo mechanical operations and are resistant to various electroplating and etching processes. Also referred to as dry film photoresist.

drying time A portion of the reflow process after preheat and before the reflow spike where volatile materials escape from the solder paste.

dual fixture A test fixture with two separate bed-of-nails units to speed up production testing. One board can be loaded on one fixture while another board is being tested on the other.

dual solder wave A wave soldering process wherein the first wave, a multi-directional, vertical jet, lavishes solder on all the surfaces it contacts. The second wave, which is laminar and flat, disperses a finishing solder to remove bridges and icicles. This process was designed for surface mount soldering. Also referred to as dual-wave soldering.

dummy component A mechanical package, typically without the silicon die, used to verify manufacturing processes inexpensively.

dummy land A conductor on a PCB that is not electrically connected to the rest of the circuitry. Frequently used as a way to reduce the gap between a component and board when using adhesive to mount the component onto the board in wave soldering or single pass reflow soldering processes. Also known as dummy track.

dynamic RAM (DRAM) Read/write memory that must be refreshed (read or written into) periodically to maintain the storage of information.

edge connector A one-piece receptacle containing female contacts, designed to receive the edge of a PCB and interconnect, on which the male contacts are etched or printed. The connector may contain either a single or double row of female contacts. Both thermoplastic and thermosetting insulating materials are used.

edge definition The fidelity of reproduction of a pattern edge relative to the production master.

electroless copper deposition Process in which base laminate is coated with a layer of copper through chemical deposition.

electromagnetic compatibility Principle in which any electronic or electrical appliance should be able to operate without causing electromagnetic interference, and without being affected by electromagnetic interference. Also known as EMC.

electrostatic discharge (ESD) A transfer of electrostatic charge between bodies at different electrostatic potentials caused by direct contact or induced by an electrostatic field.

embedding Encapsulation process using thick protective material.

emissivity The ratio of the emissive power of a body to the emissive power of a blackbody at the same temperature.

emulsion (1) A material such as polyvinyl acetate that is built up on a printing screen to block portions of the screen. The open portions define the pattern for depositing solder paste on a PCB. (2) The light-sensitive silver halide coating of photographic films that produces an image after exposure and development.

encapsulation Term referring to the protection of assemblies, reducing the effects of environmental conditions. (See Figure 8 on following page.)

entrapment The damaging admission and trapping of air, flux, and fumes. It is caused by contamination and plating.

environmental stress screening (ESS) Manufacturing stage in which all assemblies are subjected to abnormal stresses, with the aim of forcing all early failures to occur. Also known as reliability testing.

environmental test A test or series of tests used to determine the sum of external influences affecting the structural, mechanical, and functional integrity of any given package or assembly.

epoxy (1) A polymer thermosetting resin able to form a chemical bond to plastic and metal surfaces. (2) A polymeric resin material. A family of thermosetting resins used in the packaging of semiconductor devices that form a chemical bond to many metal surfaces.

epoxy smear Epoxy resin that has been deposited onto the surface or edges of the conductive pattern during drilling. Also called resin smear.

etch factor The ratio of the depth of etch (conductor thickness) to the amount of lateral etch (undercut).

etchant Solution used to remove unwanted copper from non-circuit areas on printed circuit boards.

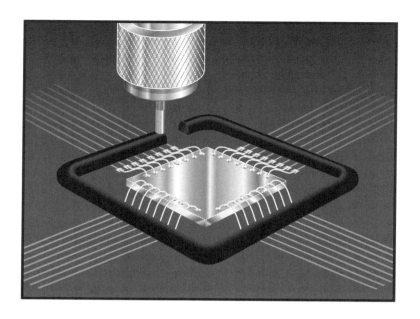

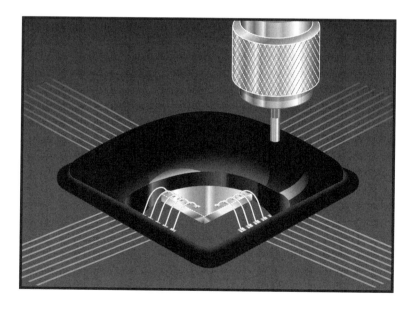

Figure 8 Encapsulation of direct chip attach (DCA); in this example, chip-on-board (COB). (Courtesy of Camelot Systems, Inc.)

etchback The controlled removal of all components of base material by a chemical process on the side wall of holes in order to expose additional internal conductor areas.

etching The process of removing unwanted metallic substance bonded to a base via chemical or chemical and electrolytic means.

etching resists Materials deposited on the surface of a copper-clad base material that prevent the removal by etching of the conductive areas they cover.

eutectic The alloy of two or more metals that has the lowest melting point. Eutectic alloys, when heated, transform directly from a solid to a liquid without experiencing a plastic region.

fabrication The bare board manufacturing process beginning after design and prior to assembly. Specifically, printed circuit fabrication relates to the process of producing one or more layers of conductive paths, or circuits, on a layer or layers of dielectric, rigid, or flexible substrate material. In printed circuit board applications, epoxy/glass or polyimide materials usually are substrates of choice; however, other materials are successfully used. Included within the broad definition of printed circuit board fabrication are a variety of processes, including lamination of conductive and insulative layers; addition or subtraction of metal, usually copper, to form conductive circuit paths in the configuration of the artwork design; drilling; application of plating masks; plating through holes; routing and profiling; and cleaning.

face bonding The opposite of back bonding. A face bonded semiconductor chip is one that has its circuitry side facing the substrate. Flip-chip and beam lead bonding are the two common face bonding methods.

failure Termination of a device or a system's ability to perform its function.

fault list A list that reflects shorts and opens on a given bare PCB after analysis. This list can be in either X-Y coordinates, test point position, and/or component position.

fault profile A description of the type and frequency of electrical faults most likely to be found on the populated PCB assembly.

feedthrough (1) A conductor that connects patterns on opposite sides of a PCB. Also called interfacial connection. (2) A connector or terminal block, usually having double-ended terminals, which permits simple distribution and bussing of electrical circuits.

fiducial A specific mark typically incorporated in the artwork and etched along with the circuitry of the PCB. The mark is used by machine vision systems to identify the artwork orientation and location on the PCB. Both global fiducials and local fiducials can be used. Global fiducials locate the overall circuitry pattern to the PCB, whereas local fiducials are used at component locations, typically fine pitch patterns, to increase the placement accuracy.

fillet (1) A radius or curvature imparted to inside meeting surfaces. (2) The concave junction formed by the solder between the footprint pad and the SMC lead or pad.

fine lines Term referring to accurate PCB production with very narrow track widths.

fine pitch The center to center lead distance of surface mount packages that are between 0.025" to 0.032".

fine pitch technology (FPT) A surface mount assembly technique with component leads on less than 0.035" centers.

finish Protection of the surface of an object.

finite-element analysis A computer-based method that divides geometric entities into smaller elements and links a series of equations to each so that they can be evaluated together.

first article A part or assembly manufactured before the start of a production run to determine whether processes are capable of fulfilling end-product requirements.

first pass yield The number of finished units in a group that pass all tests without rework.

fixture A device that provides the actuating mechanism to press the product being tested into the spring contact probe test pattern. It contains either a dedicated head or an interface for interchangeable test heads and a means of locating the product to be tested. Also called handler, adaptor, and bed-of-nails.

flare The unwanted enlargement of the area around a punched hole on the exit side of the punched or drilled PCB.

flash point Temperature at which a volatile liquid mixes with air in proportions that produce a flammable gaseous mixture.

flat pack An integrated circuit package that has leads extending from the package in the same plane as the package.

flex-rigid printed circuit A rigid PCB with a flexible portion.

flexible printed circuit A random arrangement of printed wiring and components utilizing flexible base materials with or without flexible cover layers.

flip-chip A leadless, monolithic structure, usually a silicon die, containing circuit elements, which is designed to electrically and mechanically interconnect to the circuit by means of conductive bumps, usually solder, located on its face.

flood bar A bar or other such device on a screen printing system that will drag solder paste back to the starting point after the squeegee has made a printing stroke. The flood stroke returns the paste without pushing it through the meshes, so it does no printing, only returns the paste supply to be ready for the next print.

fluorocarbon Compound of fluorine and carbon.

flux A material used to promote fusion or joining of metals in soldering, welding, or smelting. A wide range of rosin fluxes are available for soldering electrical/electronic components.

flux activity A measure of the cleaning or scrubbing ability of a flux in a soldering process.

flux residue Particles of flux materials remaining on the PCB after soldering and cooling.

fluxing The process of removing oxides from the PCB to prepare it for soldering. Fluxing lowers the surface tension of the solder and helps promote the formation of good solder fillets.

foam fluxing A type of flux transformed from a liquid to a foaming state via a porous diffuser for cleaning solder joints prior to the wave.

foil A thin, continuous sheet of metal, usually copper or aluminum, used as the conductor for printed circuits. Foils for printed circuits are commonly one ounce per square foot. The thinner the foil, the lower the required etch time. Thinner foils also permit finer definition and spacing.

footprint (1) A group of metal pads on the PCB arranged in the same pattern as the leads or pads on an SMC and to which the leads or pads are soldered. (2) The area needed on a substrate for a component or element. Usually refers to specific geometric pattern of a chip. Also referred to as land pattern.

free surface energy Energy on a material's surface that is available for molecular bonding.

fugazi A capacitance discharge unit retaining shield.

full liquidus temperature The temperature at which the molten solder reaches its maximum liquidicity and best viscosity for good wetting, typically 15–20°C above the solder alloy melting point.

functional test An electrical test of an entire assembly that simulates the intended operating environment of the product.

functional tester An automatic piece of test equipment that tests a finished, populated PCB assembly as a complete, functional entity, typically by applying inputs and sensing outputs only through its edge connector.

functional testing Electrical testing of an entire assembly to simulate an actual operating environment.

fusing Heating an electroplated tin-lead layer to ensure an interfacing alloy bond occurs.

gap Spaces between adjacent PCB tracks.

Gerber data Tabular information on aperture selection and operating commands based on dimensions in X and Y coordinates.

glass transition temperature The temperature at which an amorphous polymer (or the amorphous regions in a partially crystalline polymer) changes from a hard and relatively brittle condition to a viscous or rubbery condition.

glob-top Protective encapsulation over semiconductor die mounted directly on circuit board.

go/no-go test An electrical test of an entire assembly that simulates the intended operating environment of the product. Similar to functional test.

golden board Term used to describe ideal properties of a printed circuit board.

golden boy A term used to describe a die, component, or assembly that has been tested and is known to function to specification. It is then used in testing other units via comparison against this "perfect" or "golden boy" unit.

goober A term used to describe an ill-defined deposit of material (solder, adhesive, encapsulant, etc.) on a circuit board assembly.

gravity water separator A system that uses gravity to separate water from a hydrocarbon solvent by having the water float to the top surface of the solvent, which can then be siphoned off with a drainage system.

grid An orthogonal network of two sets of parallel equidistant lines used for locating points on a PCB on a CAD system. Connections should be located on the crosspoints of the gridlines. The position of conductors may be independent of the grid, i.e., not necessarily following the gridlines.

ground plane A conductor layer, or portion of a conductor layer, used as a common reference point for circuit returns, shielding, or heat sinking.

ground plane clearance The etched portion of a ground plane around a plated-through or non-plated-through hole that isolates the plane from the hole.

guided probe method A technique used to test production volume PCBs. It is based on the theory that proper incoming inspection will catch the majority of device failures, and that thorough inspection of the PCB prior to loading will eliminate most manufacturing errors.

gull wing lead A surface mount device lead configuration that extends outward from the body of the device.

halide-free flux Flux that does not contain halides.

halides A compound containing fluorine, chlorine, bromine, iodine, or astatine. In soldering, these materials may be part of the flux system, mostly the activator. The residues of ionic halides are considered dangerous, and must be washed.

halo effect Minor delamination of substrate material, resist, or laminate around holes.

halogenated hydrocarbon An organic compound in which some, or all, of the hydrogen atoms linked to the carbon atoms are replaced by atoms of the halogen family. One of the largest uses for these compounds is as solvents.

halogens A group of elements similar in their properties and chemical activities. These elements, in order of decreasing activity, are fluorine, chlorine, bromine, and iodine.

haloing A condition existing in the base laminate of a PCB in the form of a light area around plated through holes or other machined areas on or below the surface of the laminate. It is localized delamination.

hard solder A high melting solder with a melting point above 1100°F.

hard water Water containing calcium carbonate, and other ions such as magnesium and iron that tend to form insoluble deposits. These minerals may collect on the inside of cleaning equipment, causing a scale that can clog piping.

hardener   A catalyst added to a thermoset plastic to hasten curing. Also referred to as a curing agent.

hardware   Shells, guide pins, polarizing pins, strain relief clamps, mounting screws, etc. that are attached to the PCB.

heat and pull   Desoldering method using a soldering iron with a heater block attached, a solder pot or solder fountain, or a device that grasps, heats, and pulls the component leads to be removed.

heat management   PCB design philosophy that ensures adequate heat dissipation. Also known as thermal design.

heat sink   A method of conducting heat away from an area, usually a component die. May be internally or externally mounted to a package.

heated collet   A method of soldering surface-mounted components using an electrically heated collet positioned over the component terminals so that solder under the terminals melts.

heaters, IR panels   Radiant energy emitters that emit IR energy in the middle to far infrared region from a plane surface.

heel crack   A crack in the solder fillet across the lead bond width in the heel area.

hermetic   Sealing to be gas tight. The test for hermeticity is to observe leak rates when the object is filled with a gas, typically helium, and placed in a vacuum. A plastic encapsulation cannot be hermetic by definition because it allows gases to permeate.

hi rel A contraction of high reliability. Refers to products that are assembled and inspected under rigid standards, given extra testing and conditioning, and typically used in a military, space, or medical application.

high speed digital functional testing A test of microprocessor-based boards that incorporates a data compression method called signature analysis. The board itself can be used running at speed to provide the test stimulus. Each device is exercised and its response, or signature, is then compared to a known good board. This test method is most effective for bus oriented boards.

high reliability soldering A soldering technique whereby the probability of obtaining perfect metallic joining, product cleanliness, and optimum electrical conductivity without damage to components or equipment has been statistically proven.

high voltage testing Testing conducted at 40V or more.

hole breakout A condition in which a hole is not completely surrounded by the land.

hot air (solder) leveling (HASL) A process used in solder dipping of bare copper circuitry in which high-velocity air is used to blow solder clear from plated-through holes and to minimize solder thickness.

hot cracking A cracking of a metal or alloy upon freezing. In relation to solder, this can occur as a result of stresses developed in the solder joint by uneven cooling.

hot short Brittleness in a solder joint due to elevated temperature.

hot spot A small area on a circuit that is unable to dissipate the generated heat and therefore operates at an elevated temperature above the surrounding area.

hot tinning The application of a solderable coating by a molten solder/wetting process. Tinning is not limited to the application of pure tin, but can be achieved with any solder alloy.

hot vapor soldering Reflow soldering method using condensation inert heating as the heat transfer medium.

hot zone The part of a continuous furnace or kiln that is held at maximum temperature. Other zones are the preheat zone and cooling zone.

hybrid circuit A microcircuit consisting of elements that are fabricated directly on the substrate material in combination with discrete add-on components. Also known as hybrid assembly.

hydrocarbon Organic chemical compounds containing only hydrogen and carbon atoms.

hydroscopic A material that is capable of absorbing moisture from the air.

At Butterworth-Heinemann, we are dedicated to providing you with quality service. So that we may keep you informed about titles relevant to your field of interest, please fill in the information below and return this postage-paid reply card. Thank you for your help, and we look forward to hearing from you!

What title have you purchased? ⌊__|__|__|__|__|__|__|__|__|__|__|__|__|__|__|__|__|__|__|__|__|__|__|__|__|__|__|__|__|__⌋

Where was the purchase made? ⌊__|__|__|__|__|__|__|__|__|__|__|__|__|__|__|__|__|__|__|__|__|__|__|__|__|__|__|__|__|__⌋

Name ⌊__|__|__|__|__|__|__|__|__|__|__|__|__|__|__|__|__|__|__|__|__|__|__|__|__|__|__|__|__|__|__|__|__|__|__⌋

Job Title ⌊__|__|__|__|__|__|__|__|__|__|__|__|__|__|__|__|__|__|__|__|__|__|__|__|__|__|__|__|__|__|__|__|__⌋

Institution ⌊__|__|__|__|__|__|__|__|__|__|__|__|__|__|__|__|__|__|__|__|__|__|__|__|__|__|__|__|__|__|__|__⌋

Address ⌊__|__|__|__|__|__|__|__|__|__|__|__|__|__|__|__|__|__|__|__|__|__|__|__|__|__|__|__|__|__|__|__|__⌋

Town/City ⌊__|__|__|__|__|__|__|__|__|__|__|__|__|__|__|__|__|__|__|__|__|__|__|__|__|__|__|__|__|__|__|__⌋

State/County ⌊__|__|__|__|__|__|__|__|__|__|__|__|__|__|__|__|__|__|__|__|__|__|__|__|__|__|__|__|__|__⌋

Zip/Postcode ⌊__|__|__|__|__|__|__|__|__|__|__|__|__|__|__|__|__|__|__|__|__|__|__|__|__|__|__|__|__|__⌋

Country ⌊__|__|__|__|__|__|__|__|__|__|__|__|__|__|__|__|__|__|__|__|__|__|__|__|__|__|__|__|__|__|__|__⌋

Telephone ⌊__|__|__|__|__|__|__|__|__|__|__|__|__|__|__|__|__|__|__|__|__|__|__|__|__|__|__|__|__|__|__⌋

email ⌊__|__|__|__|__|__|__|__|__|__|__|__|__|__|__|__|__|__|__|__|__|__|__|__|__|__|__|__|__|__|__|__|__⌋

☐ Please keep me informed about other books and information services on this and related subjects.

(FOR OFFICE USE ONLY)

BUTTERWORTH-HEINEMANN IS ON THE WEB – http://www.bh.com/

(For cards outside the US please affix a postage stamp)

NO POSTAGE
NECESSARY
IF MAILED
IN THE
UNITED STATES

BUSINESS REPLY MAIL
FIRST CLASS MAIL   PERMIT NO. 78   WOBURN, MA

POSTAGE WILL BE PAID BY ADDRESSEE

DIRECT MAIL DEPARTMENT
BUTTERWORTH-HEINEMANN
225 WILDWOOD AVE
PO BOX 4500
WOBURN MA 01888-9930

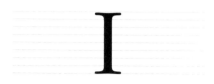

I-lead   A solder joint formed with a "footless" lead. Also referred to as a butt joint.

icicling   The formation of solder that appears like icicles resulting from poor drainoff of liquid solder following hand, wave, or dip soldering of PCB assemblies, but does not make contact with another conductor.

IEEE   Institute of Electrical and Electronics Engineers

IEEE-488 BUX   A data transmission bus that provides communication between the tester and external devices.

image fiducial   Global fiducial marks on a multiple PCB fabrication panel located within the perimeter.

impedance   The AC resistance of a circuit expressed in ohms. Impedance is frequency dependent.

impregnation   Encapsulation process where protective material is injected into all spaces or voids between components.

in-circuit test   Test procedure in which circuit nodes of an assembly are accessed by pin-type probes, to test individual components within the circuit. Also known as defects analysis and pre-screening.

in-circuit tester   An automated piece of test equipment that tests each separate device on a board by applying test signals directly to the device's inputs and sensing the results directly from the device's output.

in-circuit testing Component-by-component test to verify proper component placement and orientation and to ensure that components meet specifications.

in-line system Allows conveyorized handling of an assembly into and out of the equipment.

induction soldering A method of soldering in which the solder is reflowed or supplied by preforms. If the work is moved slowly through the energy field, the induction process may be made continuous.

infrared (IR) Part of the electromagnetic spectrum between the visible light range and the radar range.

infrared fusing The melting of tin-lead plating on PCBs using the energy given off by infrared waves.

infrared ovens Units that dry, cure, and preheat parts directly (i.e., without heating the oven air) via infrared energy.

infrared radiation The band of electromagnetic wavelengths lying between the extreme of the visible and the shortest microwaves. The strong absorption of infrared by many substances renders it a useful means of applying heat energy.

infrared soldering Soldering process using infrared radiating elements or emitters to create heat.

inner lead bonding (ILB) Process of bonding termination, which leads to a tape automated bond integrated circuit's bumps.

inorganic Refers to the chemistry of those compounds found in nature or synthetic materials that do not depend essentially upon the chemistry of carbon for their properties.

inspection facility The combination of equipment, personnel, and procedures that take inspection measurements and evaluations to establish conformance to specifications.

insufficient heat joint A solder joint formed under too low a heat or for too short a contact time. Characterized by poor wetting, poor solder rise, or a chalky appearance.

integrated circuit (IC) A small, complete circuit made by vacuum deposition and other techniques, usually on a silicon chip, and mounted in a package.

interconnection Mechanically joining devices together to complete an electrical circuit.

interconnectivity A measure of the number of solder joints that can be realized per unit area of PCB. It depends on the component technologies used and the minimum conductor geometries of the PCB.

interlayer connection An electrical connection between conductive patterns in different layers of a multi-layer PCB.

intermetallic compound (IMC) Additional quantities of the same solute that form a new crystalline phase with accurate stoichiometric proportions once a metallic solution becomes saturated. These compounds also may form in interfacial layers and during solid state diffusion. IMCs formed in the tin-lead solder system are hard and brittle, an undesirable result. IMCs affect the quality of a bond and are usually minimized for long term reliability.

internal layer A conductive pattern contained entirely within a multi-layer PCB.

interstitial via hole A plated-through hole connecting two or more conductor layers of a multi-layer PCB but not extending fully through all the layers of base material comprising the board.

ionic cleanliness Degree of surface freedom from contamination with respect to the number of ions or weight of ionic matter per unit square of surface.

ionograph An instrument designed to measure the amounts of ions present on a surface. It extracts ionizable materials from the surfaces of the part to be measured, and records the rate of extraction and the quantity.

ISHM International Society for Hybrid Microelectronics

ISO International Standards Organization

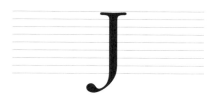

J-lead A surface mount device lead that is formed into a shape resembling the letter "J" because the lead bends underneath the device body.

JEDEC Joint Electronic Device Engineering Council

joint The connecting point of two conductors.

jumper or jumper wire An electrical connection between two points on a printed board, added after the board is fabricated.

junction A joining of two different semiconductors or of semiconductors and a metal. The four junction types include alloy, diffused, electrochemical, and grown.

kari butanol value The measure of the strength of a solvent, i.e., its ability to dissolve soils. The higher the value, the more effective the solvent but also the greater its tendency to attack delicate plastics.

kiss pressure Initial pressure applied to layers to be bonded into multi-layer printed circuit boards whereby the prepreg layers soften and flow to fill voids within the layers.

known good board (KGB) A correctly operating PCB. It is used in learning or debugging a test program in development and for comparison testers where it serves as the standard unit by which other PCBs are compared.

known good die The semiconductor dice that have been tested and are known to function to specification. Also, a die used in testing other dice.

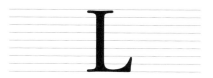

laminar flow A constant and directional flow of filtered air across a clean workbench. The flow is usually parallel to the surface of the bench.

laminar wave A smoothly flowing solder wave with no turbulence. Used in a dual wave soldering system to minimize the formation of solder bridges and icicles.

laminate The raw material for printed circuits consisting of a sheet of plastic with copper foil adhered to one or both sides.

land On a printed circuit, it is the conductive area to which components or separate circuits are attached. A plated-through hole land is the conductor surrounding the hole through the conductive pattern and the base material. Also called boss, pad, terminal point, tab, spot, or donut.

land pattern (1) A group of metal pads on the PCB arranged in the same pattern as the leads or pads on an SMC and to which the leads or pads are soldered. (2) The area needed on a substrate for a component or element. Usually refers to specific geometric pattern of a chip. Also called the footprint.

lap joint Solder joint between two metal surfaces where surfaces overlap.

large scale integration (LSI) Arrays of integrated circuits on a single substrate that comprise 100 or more individual active circuit functions or gates.

laser reflow soldering Focused heat energy from an LED, CO_2, Nd:YAG, etc. source that is directed to specific points on the PCB to reflow joints very quickly.

lay-up The technique of registering and stacking layers of a multi-layer board before the laminating cycle.

layer One of several films in a multiple film structure on a substrate or PCB.

layer-to-layer spacing The thickness of a dielectric material between adjacent layers of conductive circuitry in a multi-layer PCB.

layout Original design of a circuit board. Includes the circuitry, locating marks, pilot holes, identification marks, and number of units per board.

leaching In soldering, the dissolving (alloying) of the material to be soldered into the molten solder. Also known as scavenging.

lead The conductor brought out from a component or circuit.

lead configuration The conductors extending from a surface mount package and that function as both the mechanical and electrical connection when formed to a particular configuration. Types of lead configurations include gull wing, J-lead, and I-lead formats.

lead frame Integrated circuits are connected to lead frames to facilitate making connections to and from the various solid-state devices of the package. Typical lead frame materials include Kovar, nickel, Alloy 42, and copper.

lead pitch The center-to-center distance between leads of a component package, expressed in mils or mm. Also referred to simply as pitch.

leaded component A component with wire terminations.

leadless ceramic chip carrier (LCCC) A package in which an integrated circuit chip can be mounted to form a surface mounted component. It is made of a ceramic material that can withstand high temperatures and can be hermetically sealed. It does not have leads but has pads around its perimeters (called castellations) for connection to the PCB footprint.

leadless device A chip device having no input/output leads. Also known as leadless component.

leakage Loss of insulation between conductors on a board. May be due to improper cleaning procedures that leave conductive residues.

legend Letters, numbers, symbols, and/or patterns on the PCB that are used to identify component locations and orientation for aid in assembly and rework/repair operations.

leveling A term describing the settling or smoothing out of the screen mesh marks in thick films that takes place after a pattern is screen printed.

levels of interconnection (1) Device-to-board or -chassis: The connection point between components and the PCB or chassis. (2) Board-to-motherboard or -backplane: The connection point between PCBs or sub-circuit modules and the motherboard or a backplane board. (3) Backplane wiring: Connections between levels to each other and to other subcircuits. (4) Input/output: Connections for power and signals into and out of a system. Connections may be subassemblies within the same enclosure or between individual units.

line A single conductor on a PCB.

line certification Certification that a production-line process sequence is under control and will produce reliable circuits in compliance with requirements of applicable mandatory documents.

line width The width of a conductor on a PCB. Etched conductors are measured at their narrowest cross section.

liquidicity The degree of liquidness or viscosity of a molten solder alloy.

liquidus The temperature at which an alloy is completely molten.

loaded board Component, printed circuit board, or assembly subjected to a test. Also known as unit under test (UUT) and device under test (DUT).

loading factor Percentage of conveyor area that is covered by product.

local fiducial A fiducial mark to locate the position of land patterns for components on a PCB.

low voltage testing Testing conducted at 10V.

Manhattan effect A soldering defect in which a chip component has been pulled into a vertical or near vertical position with only one termination soldered to the PCB. It is typically caused by force imbalances during the reflow soldering process. Also referred to as drawbridging, tombstoning, and the stonehenge effect.

manufacturing defects analysis Test procedure in which circuit nodes of an assembly are accessed by pin-type probes to test individual components within the circuit. Also known as in-circuit testing and pre-screening.

mask The photographic negative that serves as the master for making thick and thin film screens.

mass soldering Process which solders many components to a printed circuit board simultaneously.

master drawing A document showing dimensional limits or grid locations for an assembly to be fabricated. Includes the arrangement of conductors and non-conductive patterns; the size, type, and locations of holes; and other information.

mealing A condition that occurs between a conformal coating and the base material, resulting in spots or patches that reveals a separation of the conformal coat from the PCB surface and/or components.

mean time between failure (MTBF) The arithmetic or statistical mean average time interval, usually in hours, that may be expected between failures of an operating unit. Results should be designated actual, predicted, or calculated.

mean time to failure (MTTF) Average time between failures.

mean time to repair (MTTR) A measure of how long it takes to access a failed system and identify, locate, and repair the fault.

measling A condition existing in the base laminate of printed circuits in the form of discrete white spots or "crosses" below the surface of the base laminate, reflecting a separation of fibers in the cloth at the weave intersection.

metal electrode leadless face (MELF) A round, cylindrical passive component with a metallic cap termination located at each end.

melting point The temperature at which a solid material becomes liquid. For metals there is a sharp melting point only in the case of a pure element, or a eutectic alloy.

meniscograph An instrument used to measure the solderability of surfaces, using the wetting balance method. It measures and plots the time and forces required for the test specimen to change from buoyancy in the solder to a downward (wetting) pull.

mesh size Number of holes per linear measure in a screen-printing material.

metal cored PCB A printed circuit board with an internal thermal plane.

metallization A metallic material deposited to enable electrical and mechanical interconnections. It may be deposited on substrates, component terminations, or the base metal.

microcircuit The physical realization of a hybrid or monolithic interconnected array of very small active and passive electronic elements.

mil A unit of length equal to 0.001".

MIL-STD-883 The basic military standard for reliable semiconductors.

Mil. Std. Abbreviation for Military Standard—the basic military standard for reliability.

military grade IC Typically, an IC whose performance is guaranteed over the temperature range from −55°C to +125°C.

military specifications Documents issued by the Department of Defense that define materials, products, or services used only or predominantly by military activities.

minimum angular ring The minimum width of metal, at the narrowest point, between the edge of the hole and the outer edge of the terminal area. The measurement is made to the drilled hole on internal layers of multi-layer PCBs and to the edge of the plating on outside layers of multi-layer boards and double-sided boards.

misalignment Passive surface mounted components that have moved during the soldering process.

mixed assembly Electronics assembly in which through-hole and surface mount components are inserted and mounted. Also referred to as mixed technology assembly.

molded circuit board A printed circuit board comprised of an injected molded base and plated track. Also referred to as three-dimensional circuit board and molded PCB.

monolithic ceramic capacitor A term sometimes used to indicate a multi-layer ceramic capacitor.

most vulnerable component (MVC) The component on a circuit board assembly with the lowest thermal threshold of pain.

motherboard A PCB used for interconnecting arrays of plug-in electronic modules.

mounting hole A hole used for the mechanical mounting of a printed board or for the mechanical attachment of components to the printed board.

moving probe Test fixture method in which two or more probes are robotically controlled to move among multiple test points on a printed circuit board.

muffle A rectangular cross section enclosure, located between the heating elements and the parts being processed, which contains the atmosphere required for the reflow process.

multi-layer board (MLB) A printed circuit board that has more than two metallic patterns in the X and Y direction. The layers are interconnected by the plated-through holes.

multi-layer printed board A multi-layer board formed by sequentially laminating through hole plated double sided or multi-layer boards together. The circuitry layers are interconnected with interstitial via holes and through connections.

multi-layer printed circuit A printed circuit board with alternate layers of conductors and dielectric, bonded together and interconnected with plated through holes or posts.

multi-level stencil A stencil of varying thickness to accommodate coarse and fine-pitch deposits. Also known as step-down stencil.

multi-wired printed circuit A printed circuit board, comprising a single- or double-sided board, with insulated wire tracks laid on to build up connections.

nails Spring-loaded metal probes used in a bed-of-nails fixture to make electrical contact with the nodes on a circuit board.

negative An artwork, artwork master, or production master in which the intended conductive pattern is transparent to light, and the areas to be free of conductive material are opaque.

net list (1) A list of designations of groups of two or more points in a circuit design that are electrically common. (2) An analyzer-generated list that defines those points that are common to one trace/network and independent points.

neutralizer An alkaline chemical added to water to improve its ability to dissolve organic acid flux residues.

nickel A metal that offers a combination of corrosion resistance, formability, and tough physical properties.

nitrogen containment capability Ability of an oven to contain nitrogen. Expressed in parts per million (ppm) of residual oxygen as measured anywhere in the heated process area of the oven tunnel, unless otherwise specified.

nitrogen inerted soldering Mass soldering within a nitrogen gas atmosphere to limit the effects of oxidation on the solder joints.

no-clean flux Solder flux with a low-solids residue that does not have to be cleaned from the circuit board following soldering.

no-clean soldering A soldering process that uses a specially formulated solder paste that does not require the residues to be cleaned after solder processing.

node An electrical junction that connects two or more component terminations.

non-activated flux Rosin based flux without activator.

non-ionic A substance that does not ionize (lose electrons) in water.

non-polar compound A compound that has electrical charges symmetrically distributed over the surface of the molecule, and therefore shows no electrical effects in solution or otherwise.

non-wetting A condition whereby a surface has contacted molten solder, but has had part or none of the solder adhere to it. Non-wetting is recognized by the fact that the bare base metal is visible. It is usually due to the presence of contamination on the surface to be soldered.

omegameter An instrument used to measure ionic residues on the surface of PCB assemblies. The measurement is taken by immersing the assembly into a predetermined volume of a water-alcohol mixture with a known high resistivity. The instrument records and measures the drop of resistivity due to ionic residue over a specified period of time. Also see ionograph.

onsertion Colloquial expression referring to the placement of components on the surface of a circuit board, as in surface mount technology.

open A fault that causes two electrically connected points to become separated. It is an area of a bare PCB which, due either to over-etching or fabrication related problems, interrupts the intended design on the circuit. It also may be caused by insufficient solder from wicking or gross coplanarity of the lead at the point of connection between the PCB and SMC. Also referred to as an open joint.

organic Relates to the chemistry of carbon compounds. The majority of carbon-containing compounds fall into the organic chemistry class. An example of a carbon compound that is not considered organic is carbon dioxide gas.

organic halides Organic compounds containing halogens.

organic solder preservative (OSP) Layers of organic coatings applied to entire board surfaces to prevent oxidation and to retain solderability.

ounces of copper This refers to the thickness of copper foil on the surface of the laminate: $\frac{1}{2}$ ounce copper, 1 ounce copper, and 2 ounce copper are common thicknesses. One ounce copper foil contains 1 ounce of copper per square foot of foil. The foil on the surface of laminate may be designated for the copper thickness on both sides by: $\frac{1}{1}$ = 1 ounce, two sides; $\frac{2}{2}$ = 2 ounces, two sides; and $\frac{2}{1}$ = 2 ounces on one side and 1 ounce on the other side. $\frac{1}{2}$ oz. = 0.72 mil = 0.00072"; 1 oz. = 1.44 mils = 0.00144"; 2 oz. = 2.88 mils = 0.00288".

outer lead bonding Process of bonding the terminations of a tape automated bond integrated circuit to a circuit board.

outgassing De-aeration or other gaseous emission from a PCB assembly (printed board, component, or connector) when exposed to a reduced pressure or heat, or both.

overhang (1) PCB: Increase in printed circuit conductor width caused by plating buildup or by undercutting during etching. (2) Assembly: Refers to a lead or leadless termination that extends beyond the pad of the PCB.

oxidation Oxidation can be considered in two senses. In the first and narrow sense, oxidation can be considered the simple addition of oxygen to a metal, e.g., the addition of atmospheric oxygen to iron to familiar rust. In the broader chemical sense, oxidation is any process where a metal loses electrons and is converted from metal with zero electrical charge, to a metallic ion with a positive charge.

oxygen analyzer An instrument that detects and calculates in ppm the amount of oxygen in a given volume.

oxygen inhibition An undesirable side reaction occurring when UV ink is irradiated (during UV curing processes) and the initiator breaks down into the reactive species needed to polymerize the ink binder.

P

P/I structure Packaging and interconnection structure. A generic term describing the substrate structure on which electronic components are mounted.

package The container for an electronic component with terminals to provide electrical access to the inside of the container. It usually provides hermetic and environmental protection for, and a particular form factor to, the assembly of electronic components.

packaging density The quantity of functions (components, interconnection devices, mechanical devices) per unit volume, usually expressed in qualitative terms, such as high, medium, or low.

pad (1) The portion of the conductive pattern on printed circuits designated for the mounting or attachment of components. (2) The area of a printed circuit used for making connections to the pattern; also the area around a drilled or plated-through hole.

pads only layer Foil outer layers of a PCB used to replace the screened on solder mask layers. It has pads without circuit traces connected to them.

panel plating Processes in metallization of PCB track in which tin-lead is selectively plated.

partial lift A bonded lead partially removed from the bonded area.

passivation A form of surface oxidation that acts as a barrier to further oxidation or corrosion.

paste *See* solder paste.

peel back The point in the wave soldering process when the solder in the wave breaks away from the exiting board and returns to the solder wave.

peel strength Strength of adhesion between two joined materials. Also known as bond strength.

peel-away speed The rate of stencil removal after printing. Expressed in in./sec.

percent area coverage The actual footprint area covered with paste or adhesive.

percent contribution The amount (%) that a single factor contributes to a total variation.

pH Measure of acidity or alkalinity of a solution. A pH of 7 is neutral—neither acidic or alkaline. Solutions with pH above 7 are alkaline and those below 7 are acidic.

phase diagram Graphical representation of temperature phases in an alloy. A tin-lead phase diagram (i.e., solder) shows solidus and liquidus temperatures for a range of tin-lead proportions.

phonon A unit of thermal energy in a crystal structure.

photo-printing Process of photographically applying a resist to the surface of a printed circuit board.

photomask Surface on which the master artwork is projected. It is used to expose the laminate, pre-coated with the photoresist, to ultraviolet light.

photoplotter A high accuracy ($\pm 0.002"$ or better) flatbed plotter with a programmable photo image projector assembly. It is most often used to produce actual size master patterns for printed circuit artwork directly on dimensionally stable, high contrast photographic film.

photoresist A blend of solid or liquid chemical compounds that can be applied to a panel surface as a thin film. This mixture contains a light-sensitive substance that permits an image to be transferred on to the panel.

phototool Photomask, photographic master, etc.

physical layer The total conductive pattern formed upon one side of a single layer of base material. See also conductor layer.

pick-and-place machine A programmable machine usually having a robot arm that picks up components from an automatic feeder, moves to a specified location on a PCB, and places or inserts the component onto or into the correct location. (See Figure 9 on next page.)

pickling A stage of electroplating in which the PCB surface is prepared for subsequent electroplating processes.

pilot hole The hole used to position a board for other operations so registration will be accurate.

pin grid array (PGA) Integrated circuit package in which the input and output points are through-hole pins arranged in a grid pattern.

pinhole A small hole in the solder surface.

pitch The nominal distance from center-to-center of adjacent conductors.

pits Small holes or sharp depressions in the surface of solder that may be caused by flux blown-out due to entrapment or overheating.

pitted solder joint This solder joint shows evidence of pits, pin holes, or small craters in the solder. This joint can be caused by oxidation, the type of plating material

74 placement

Figure 9 Component placement head (pick and place) holding QFP. (Courtesy Mydata Automation)

used on conductors (gold plating), or other foreign matter not compatible with solder. The joint may appear to have a dull finish, depending on the amount of contamination present. Also referred to as a porosity solder joint.

placement The device must be attached to the board in the exact location with the exact orientation required by the circuit, and this is the function of the automatic placement equipment. This equipment combines high speed and precise positioning of the devices. These placement machines can be classified into three basic types: mass transfer of surface mount devices, x-y positioning, and in-line transfer systems. These systems can be combined so that the features of the machine can match the type of device and the design of the board. Automatic placement is a major feature of SMD technology.

placement center Area of an automatic component placement machine where the component is centered to an absolute position in the placement head.

placement rate The speed of a complete component placement cycle starting from the component pick up location to the placement location, and then to the return position using actual devices.

plastic device A device wherein the package, or the encapsulant material for the semiconductor die, is plastic. Such materials as epoxies, phenolics, silicones, etc., are included.

plastic leaded chip carrier (PLCC) A package in which an integrated circuit chip can be mounted to form a surface mounted component. It is made of plastic material that can withstand high temperatures and has J-leads around its perimeter for connection to a PCB footprint.

plated through-hole A hole formed by deposition of metal on the sides of the hole and both sides of the base to provide electrical connection from the conductive pattern on one side to the opposite side of the PCB. Also referred to as PTH, through-hole, and via.

plating resist This is the resist used to image a printed circuit panel for plating or for etching. The resist may be applied by silk screen printing, or by dry film or liquid photoresist.

plenum A chamber that is used to uniformly distribute a fluid or gas (air, nitrogen, or other gas), into the process chamber.

76 polar compound

polar compound A compound in which electrical charges are distributed asymmetrically over its molecular surface. Ionizable compounds such as flux activators are usually polar compounds.

polarization The techniques of eliminating symmetry of parts so that they can be picked and placed in only one (correct) way.

polyethylene glycol An alcohol used to dissolve rosin flux residues.

polyimide film A tough, plastic film exhibiting good physical and electrical properties over a side temperature range. Produced from pyromellitic dianhydride and an aromatic diamine, it is used for multi-layer PCBs, hybrid circuit substrates, film capacitor dielectrics, and as a base for flexible circuits and tape cable.

polyimide resins High temperature thermoplastics used with glass to produce printed circuit laminates for multi-layers and other demanding circuit applications. They provide excellent resistance to frictional wear and oxidative degradations.

polymerization The process of uniting chemically two or more monomers (molecules) or polymers of the same kind to form one of greater molecular weight.

poofter Large, palm-activated switch used on production machines as emergency stop and electrical power off switch.

popcorning A condition in which microfissures are occurring in the thermoset plastic body of multi-leaded devices because the entrapped moisture is vaporizing.

porosity solder joint This solder joint shows evidence of pits, pin holes, or small craters in the solder. This joint can be caused by oxidation, the type of plating material used on conductors (gold plating), or other foreign matter not compatible with solder. The joint may appear to have a dull finish, depending on the amount of contamination present. Also referred to as a pitted solder joint.

positive (1) An artwork, artwork master, or production master in which the intended conductive pattern is opaque to light, and the areas intended to be free from conductive material are transparent. (2) Clear and opaque areas in the same order as the original. Positive phototools have black lines (circuits) on a transparent background and are used with photoresist to pattern plate.

potting Encapsulation process in which the assembly is embedded inside a container.

potting compound A compound, typically electrically nonconductive, used to encapsulate or as a filler between components, conductors, or assemblies.

preflow The portion of the reflow soldering profile, prior to the solder reaching its melting point, where flux activators commence "scrubbing" and solder paste solvents are driven off. Also known as soak and preflow soak.

preforming Act of shaping component leads prior to insertion into a printed circuit board. Also known as prepping.

preheat A composite term generally referring to a process portion of the reflow heat curve where the product is heated from ambient, at a determined rate until it reaches the preheat temperature.

preheat force That portion of the force profile where light contact is made between the hot bars and the component leads to allow for wetting of the joining material (solder) prior to application of full bonding force.

preparation fluid A material used to promote fusion or joining of metals in soldering, welding, or smelting. A wide range of rosin fluxes are available for soldering electrical/electronic components. Also known as flux.

prepping Act of shaping component leads prior to insertion into a printed circuit board. Also known as preforming.

prepreg Sheet material consisting of the base material impregnated with a synthetic resin such as an epoxy or polyimide partially cured to the B-stage. They are molded under heat and pressure for multi-layer printed circuitry and are used for bonding together the individual circuit layers or multi-layer PCBs.

preproduction test board A test board used to determine whether, prior to the production of finished boards, the contractor can produce a multi-layer board satisfactorily.

prescreening Test procedure in which circuit nodes of an assembly are accessed by pin-type probes, to test individual components within the circuit. Also known as defects analysis and in-circuit testing.

pretinning The coating of a component's leads or pads or the pads of a PCB with a material to improve solderability during assembly.

primary side The side of the PCB on which most of the components or the active components are mounted. See also component side.

print and etch A process for obtaining conductive patterns by the selective removal of unwanted portions of a conductive foil.

print head The assembly in a printer that houses and controls the squeegee blade(s).

print speed The rate at which the squeegee blade moves across the stencil during the print cycle. Typical speeds range between 0.4 to 6.0"/sec. Also known as squeegee speed.

printed circuit assembly (PCA) The generic term for a PCB after all electrical components have been attached. Also referred to as a printed wiring assembly (PWA).

printed circuit board (PCB) A part manufactured from a rigid base material upon which a completely processed printed circuit has been formed.

printed wiring assembly (PWA) The generic term for a PWB after all electrical components have been attached. See also printed circuit assembly.

printed wiring board (PWB) A substrate of epoxy glass, clad material or other material upon which a pattern of conductive traces is formed to interconnect the components that will be mounted upon it.

probe The mechanical contact by which electrical contact is made between the bare PCB and the continuity tester.

probing systems Equipment used for high-reliability testing of printed circuits, components, and assemblies. Probing devices range from manual probes for laboratory or low volume use, to computer-controlled systems.

production master A 1-to-1 scale pattern to produce a PCB within the accuracy specified on the master drawing.

profile A graph of time vs. temperature.

profiling The act of preparing the temperature levels (vs. time and travel) to be imposed on PCBs in the reflow soldering system.

protective coating Coating applied to a manufactured printed circuit board prior to assembly with components.

quadpack A generic term for a surface mount device with leads on all four sides. The most common lead configuration of such devices are gull-wing.

qualification agency The organization that is used to perform documentation reviews and audits of an inspection or testing facility.

qualified products list (QPL) A listing of manufacturers qualified by test and performance verification to produce items listed in the MIL specs.

quality Achievement of a system to conform to its specified performance.

quality conformance circuitry area A test board made as an integral part of the multi-layer printed board panel on which electrical and environmental tests may be made for evaluation without destroying the basic board.

quartz lamp Rapid-responding tungsten filament emitter used as an infrared heat source. Lamps generate IR at wavelengths of 2.5–5.0 micron in SMT applications.

quench Rapidly cooling molten solder below its melting point in order to solidify it and obtain a strong and reliable solder joint.

radial lead A lead extending out of the side of a component rather than from the end. The opposite lead configuration is the axial lead.

radiant heat transfer The electromagnetic radiation emitted by one body due to the temperature difference between it and another body, the radiation being proportional to the difference.

radiation, focused infrared Infrared concentrated on a point or line by a backup reflector.

radiation, infrared Thermal radiation emitted in the infrared wavelength region (0.7–1000 microns) of the electromagnetic spectrum.

radiation, long wave IR Infrared occurring between the wavelengths of 5–100 microns.

radiation, medium wave IR Infrared occurring between the wavelengths of 2.5–5 microns.

radiation, near IR Infrared occurring between the wavelengths of 0.78–2.5 microns. Also called short wave IR radiation.

radiation, non-focused infrared Infrared scattered over an area by a diffuse backup reflector.

radiation, re-emitted infrared Thermal energy absorbed by a media that is re-emitted as infrared at a wavelength dependent on its temperature.

radiation, reflected infrared IR energy that is redirected to a target. No change in wavelength occurs.

radiation, short wave IR Infrared occurring between the wavelengths of 0.78–2.5 microns. Also called near IR radiation.

radiation, thermal infrared Thermal electromagnetic radiation heat transfer occurring between the wavelengths of 0.78-1000 microns.

random access memory (RAM) A type of memory that offers access to storage locations within it by means of X and Y coordinates.

read only memory (ROM) A random access storage in which the data pattern is unchangeable after manufacture.

real estate The surface area of an integrated circuit or of a substrate. The surface area required for a component or element.

reduction Removal of oxygen from a compound. Reduction typically signifies a decrease in an element's or an ion's positive charge.

reference designators Diagram markings that distinguish one graphic symbol from another and correlate these identifications with actual components on the parts lists and assembly drawings. They consist of a combination of letters and numbers that identify the class of the component.

reflectivity Percentage of incident infrared that reflects from the surface, thus having no heating effect.

reflow of through-hole Process in which through-hole components populating a PCB assembly are reflow soldered simultaneously with surface mount components (rather than wave soldered or hand soldered). Also known as single center reflow soldering, paste-in-pin, pin-in-paste, and intrusive soldering.

reflow process A general term referring to the overall process of reflowing solder paste in attaching surface mount components to various substrates. It usually includes the preheat process, stabilization and/or drying, the reflow spike, and cooldown, but sometimes refers to the reflow spike area only. Also called reflow soldering.

reflow soldering A process for joining parts to a substrate by depositing solder paste, placing parts, heating until the solder fuses, and allowing them to cool in the joined position. Also referred to as the reflow process.

reflow spike A portion of the reflow process where the temperature is raised sufficiently to cause the solder paste to reflow.

registration The location of a circuit with respect to fixed points. Successive operations must register properly so opposite sides of the circuit will mate properly, holes will fall in center of pads, and tabs will fit into connectors.

repair The act of restoring the functional capability of a defective part without necessarily restoring appearance, interchangeability, and uniformity.

repeatability The ability to repeatedly return to a specific target. Used when evaluating the consistency of processes and process equipment.

resin A solid or semi-solid organic compound lacing a crystalline structure. Resins are characterized by not having definite and sharp melting points, are usually not conductors of electricity, and many are transparent or translucent. Natural resins usually originate in plants, such as pine sap, and are not water soluble. The rosin used in soldering fluxes is an example of resin.

resin recession A void between the plated through-hole barrel and the wall of the holes when viewed in cross section after exposure to high temperatures.

resin smear A condition usually caused by drilling in which the resin is transferred from the base material to the wall of a drilled hole covering the exposed edge of the conductive pattern.

resist A material used to mask some process action from a given part of an assembly. An example is a resist used in soldering to cover those parts of the bare conductor over which a solder coat is not desired.

resistance A measure of the difficulty of moving an electrical current through a material when a voltage is applied.

resistance heating A method of heating that depends on one of two principles: (1) The passage of large currents through a poor conductor like graphite or resistive alloys. Here the heat generated in the tool is transferred by conduction to the work—common in soldering. (2) The resistance of an air gap that causes an arc and rapid heating until the gap is filled by molten metal—common in brazing and welding.

resolution (1) *PCB Fab:* The ability to reproduce artwork of various size lines and spaces. (2) *PNP* (Pick and Place): Defines the finest increments of machine movement, and hence, the ultimate precision of the equipment. (3) *Vision:* The number of pixels per screen image. The larger the number, the better the resolution.

response time The time required for thermal equilibrium to occur after a setpoint change is made on a reflow machine.

rework The act of repeating one or more manufacturing operations for the purpose of bringing a non-conforming item into conformance with a drawing, specification, or contract requirement.

rheology (1) A term used to describe the flow of a fluid. Typically used to describe the viscosity and surface tension properties of a solder paste. (2) Describes the flow of a liquid, as with solder paste or epoxy, or its viscosity and surface tension properties.

robber An area designed around a printed circuit board to absorb unevenly distributed current.

rosin A naturally occurring resin usually associated as a component of pine sap. It is a mixture of several organic acids, of which abietic acid is the chief component. Available as gum, wood, and Tall Oil Rosins, sometimes chemically modified. The most widely used material in the manufacture of soldering fluxes for the electronic industry is water white (ww) gum rosin.

rosin flux The mildest and least effective of solder fluxes. To increase rosin flux efficiency, small amounts of organic activating agents are added. Type RMA, mildly activated rosin flux, is the flux most commonly used for electrical connections.

rotational error The angular displacement of the component axes resulting from component centering-mechanism inaccuracies and placement tool rotational precision.

router (CAD) A computer program that determines paths between interconnecting points.

saponification Reaction converting rosin into rosin soap. Post-soldering cleaning processes that allow water to be used as a cleaner as opposed to as a solvent.

saponifier A generic term for an aqueous solution of organic or inorganic bases and various additives for promoting the removal of rosin and sometimes water soluble fluxes. The removal of rosin-based fluxes is based on a chemical reaction between rosin acids and a base in the saponifier, rendering it soluble and/or dispersible in detergent solution as the rosin "soap."

saturated solution A solution in which the solvent can accept no more solute. The result of adding additional solute to the solution is the formation of the solute as a distinct phase, e.g., solid particles suspended or precipitated to the bottom of the container in which the solution is held.

scan rate The pace at which a machine scans the surface of a unit, such as a solder joint, to determine volume. Expressed as surface area per unit of time.

scavenged air Air removed from appropriate parts of the process area, i.e., tunnels, to ensure there are no fluid or gas vapors in the work place.

scavenging In soldering, the dissolving (alloying) of the material to be soldered into the molten solder. Also known as leaching.

schematic diagram A drawing showing, by means of graphic symbols, the electrical connections, components, and functions of a specific circuit arrangement.

screen A network of metal or fabric strands, mounted snugly on a frame, upon which the film circuit patterns and configurations are superimposed by photographic means.

screen printing A process for transferring an image to a surface by forcing suitable media through a stencil screen with a squeegee. Also called silk screening.

screening The process whereby the desired film circuit patterns and configurations are transferred to the surface of the substrate during manufacture by forcing a material through the open areas of the screen using the wiping action of a soft squeegee.

secondary side In through-hole technology, this term refers to the soldered side of the PCB. In surface mount technology, this term refers to the secondary side, which typically is limited to passive chip components. Also known as solder side.

seeding Process in electroless copper deposition where the sensitized board is dipped into an acidic solution of palladium chloride.

selective etch Restricting the etching action on a pattern by the use of selective chemicals that attack only one of the exposed materials.

self-align Design of two mating pairs so that they will engage in the proper relative position.

semi-additive process An additive process for obtaining conductive patterns that combines an electroless metal deposition on an unclad substrate with electroplating or etching, or with both.

semi-aqueous cleaning This cleaning technique involves a solvent cleaning step, hot water rinses, and a drying cycle.

semiconductor device Any device based on either preferred conduction through a solid in one direction, as in rectifiers; or on a variation in conduction characteristic through a partially conductive material, as in a transistor.

sensitize Process in electroless copper deposition where the board is dipped into acidic stannous solution.

service loop A small portion of wire or conductor that is added to the overall length to facilitate maintenance and further servicing.

shadowing (1) A condition in infrared reflow soldering in which component bodies block the direct infrared energy from certain areas of the PCB, possibly resulting in insufficient temperatures to completely melt the solder paste. (2) In wave soldering, it refers to solder failing to wet the surface mount device leads due to another device blocking the solder flow. Also known as shadow effect.

shelf life The length of time under specified conditions that a stored material in original, unopened containers retains its usability. Also called storage life.

silicon A brittle, gray, crystalline chemical element that, in its pure state, serves as a semiconductor substrate in microelectronics.

silicone A group of semi-organic polymers that have good heat stability and are water repellent. In soldering they are considered a poison to the joining process, because of their tenacity. Silicone oil, for instance, cannot be totally cleaned off a surface with electronic grade solvents.

silk screen A screen of a closely woven silk mesh stretched over a frame and used to hold an emulsion outlining a circuit pattern and used in screen printing of film circuits. Used generically to describe any screen (stainless steel or nylon) used for screen printing.

silver chromate paper test A simple qualitative test to determine the presence of ionic halides. Usually used to check that a mildly activated flux, such as type RMA, contains no ionic halides. The Silver Nitrite Test, which is another method, serves the same purpose.

silver migration Silver is a metal prone to electromigration in which the migration can extend along the surface and tends to follow paths through fibrous materials along the surfaces.

silver nitrate test A simple qualitative test to determine the presence of ionic halides. The Silver Chromate Paper Test, which is another method, serves the same purpose.

simulate A software model that predicts a board's operation before it is built.

simultaneous placement Placement of more than one surface-mounted component onto a circuit board at one time.

single in-line package (SIP) An assembly consisting of two or more surface mount components mounted on a common substrate. The components are interconnected to each other and to pins at the substrate edge for through-hole mounting on a PCB.

single wave soldering Wave soldering process using just one solder wave.

skew Describes the misalignment of a part to its target.

slump A spreading of material (solder paste, conductive adhesive, thick film composition, etc.) after screen or stencil printing but before reflow soldering, curing, or drying. Too much slump results in a loss of definition.

small outline integrated package (SOIC) A package in which an integrated circuit chip can be mounted to form a surface mounted component. It is made of a plastic material that can withstand high temperatures and has leads formed in a gull-wing shape along its two longer sides for connection to a PCB footprint.

SMTA Surface Mount Technology Association

snap-back The return to normal of a stencil after being deflected by the squeegee moving across the surface of stencil and substrate.

snap-off distance The height the stencil is set above the board for an "off-contact" printing, which determines the amount of deflection that occurs during the snap-back action of the stencil. Snap-off distance is typically in the range of 0.003 to 0.050". On-contact printing would have a zero snap-off distance.

soak Period of time after preheat and before the reflow spike where the internal temperature differences between components are allowed to equalize. Also known as drying time, preflow, and stabilization period.

soft solder A low-melting solder, generally a lead-tin alloy, with a melting point below 600°F.

soft water Processed water where the hardness has been replaced with sodium ions. This grade of water is suitable for most electronic cleaning applications and is more economical than de-ionized (DI) grade water.

soils Foreign matter that might exist on a surface to be soldered, possibly interfering with the soldering process. It may be organic or inorganic.

solder (1) An alloy that melts at relatively low temperatures, and which is used to join or seal metals with higher melting points. Solder alloys melt over a range of temperatures; the temperature at which a solder begins to melt is the solidus, and the temperature at which it is completely molten is the liquidus. (2) A metal alloy, usually having a low melting point, used to join other metals having higher melting points than the solder. Solder is an adhesive; it wets the surfaces and forms the joint by causing molecular attraction between the solder and base metal. It also may diffuse the solder metal or alloy into the base metals or vice versa. Solders are generally classified as soft solders and hard solders. Soft solders have melting points up to approximately 600°F, whereas the melting points of the hard solders are above 1100°F. The most common examples of soft solders are the tin-lead alloys.

solder balls Very small balls of solder that separate from the main body of solder which forms the joint. Primarily caused by excessive oxides in the solder paste that inhibit solder coalescence during reflow.

solder bump Solder spheres bonded to contact areas or pads of devices and used for face-down bonding.

solder cream Another term for solder paste. It is a homogeneous combination of solder, flux, solvent, and a suspension agent for automated production of solder joints. Available with a variety of flux bases.

solder dam A dielectric composition screened across a conductor to limit molten solder from spreading further onto solderable conductors.

solder foil Solder alloys in the form of thin tape in various widths, typically ranging from $\frac{1}{4}$ to 3". Also called solder tape or solder strip.

solder fusion equipment, IR Systems that change the porous, electroplated tin-lead on a circuit into an alloy with a strong bond to the base copper. Conveyorized infrared fusing systems fuse both sides of a double-sided board simultaneously. They consist of a fluid application stage

solder leveling

for applying fusing fluid, a preheat zone, a fusing zone, and a cooling zone.

solder leveling The process of dipping PCBs into hot liquids, or exposing them to liquid waves to achieve fusion. First, flux is applied to the board by dipping or brushing. Then the board is preheated in a liquid maintained at 250°F. Next, the board is immersed in fusing liquid of 430°–500°F. Finally, it is dipped in another 250°F liquid to cool it and reduce thermal shock. Thin fused coatings can be applied.

solder marks A screening defect that is characterized by a print having jagged edges. This condition is a result of incorrect dynamic printing pressure or insufficient emulsion thickness.

solder mask A PCB technique where everything is covered with a plastic coating except the contacts to be soldered.

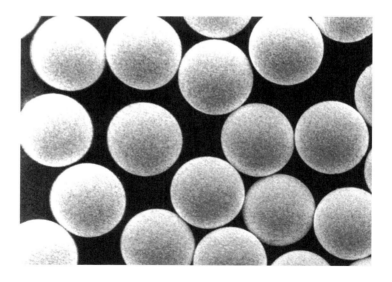

Figure 10 SEM with solder alloy spheres used in solder paste. (Courtesy MPM Corporation)

solder mask over bare copper (SMOBC) A printed wiring technology that protects bare copper conductors with solder mask exposing only the component land patterns.

solder pallets PCB holders on conveyors to the wave-solder or reflow system that are used continuously from the assembly area to eliminate excessive handling.

solder paste A homogeneous combination of solder powder, flux, solvent, and a gellin or suspension agent for automated production of solder joints. Available with a variety of flux bases.

solder preforms Manufactured solder configurations containing a predetermined quantity of alloy, with or without a flux core or coating. Available as stamped washers, spheres, and formed wire.

solder resist Permanent or temporary coatings that mask off and surface insulate those areas of a circuit where soldering is not desired or required.

solder side In through-hole technology, this term refers to the soldered side of the PCB. In surface mount technology, this terms refers to the secondary side, which typically is limited to passive chip components. Also known as secondary side.

solder skip Joint not properly soldered due to shadowing by one or more component bodies as an assembly is wave soldered.

solder strip Solder alloys in the form of thin tape in various widths, typically ranging from $\frac{1}{4}$ to 3". Also called solder foil or solder tape.

solder tapes Solder alloys in the form of thin tape in various widths, typically ranging from $\frac{1}{4}$ to 3". Also called solder foil or solder strip.

solder voids Emission of gas or air from a printed circuit board or joint as the board is soldered.

solder wicking A condition whereby solder on the terminal of a surface mounted component being soldered migrates up from the circuit board land and leaves insufficient solder on the land to give a good joint.

solderability (1) The ability of a conductor to be wetted by solder and to form a strong bond with the solder. (2) A qualitative measure of the ability of PCB pads or of a component's leads or terminations to be completely wetted by molten solder.

solderability testing There are several test systems presently in use for both PCBs and components. An additional test can be used for plated-through holes and components. They are: (1) the edge dip solderability test, (2) the meniscus test, and (3) the globule test. In addition to these, any test, mutually agreed to by vendor and customer for the solderability of components including printed circuit boards, is acceptable.

soldering The process of joining metals by fusion and solidification of an adherent alloy having a melting point below about 800°F.

soldering, reflow A process in which solder paste is deposited upon pre-tinned pads on a PCB and the component's pre-tinned leads are placed upon the paste. When the assembly is heated to the proper temperature, the solder paste melts and the solder on the leads and pads reflows to form a solder fillet.

soldering, vapor phase reflow A type of reflow soldering in which the PCB assembly is passed through a vaporized inert fluorocarbon. The latent heat given up when the fluorocarbon condenses causes the solder to reflow.

soldering, wave A method of soldering complete assemblies where the PCB with components mounted on it is passed through one or more waves of molten solder which is continuously moving to maintain fresh solder in contact with the PCB.

solid state Technology using solid semiconductors in place of vacuum tubes for amplification, rectification, and switching.

solids The percentage by weight of rosin in a flux formulation. Also see solids content.

solids content The percentage by weight of rosin in a flux formulation. Commonly referred to as "solids."

solidus The temperature at which a metal alloy begins to melt. At this temperture, some components of the alloy melt or begin to melt, whereas the balance of the material is still solid.

solute The component of a solution that is dissolved in solvent. The solute can be a solid, liquid, or gas.

solution A homogeneous mixture formed by processing, in which a solid, liquid, or gaseous substance is mixed with a liquid, solid, or gas called a solvent. The term is usually associated with liquids, but may include solids, as in alloys, or gaseous mixtures. Generally, a solution will be clear or transparent.

solvent One of the components of solution. It is that component in which the other components, i.e., the solutes are dissolved. In the case of a solution composed of several liquids, the liquid present in the greatest quantity is usually referred to as the solvent.

solvent cleaning Cleaning by means of organic solvents.

spacings The distance between adjacent conductor edges.

specific gravity　The ratio of the density of a material to the density of water.

specific gravity　The weight of a substance divided by the same volume of water at the same temperature.

specification drawings　A document of dimensional limits applicable to components and any other information germane to the product to be built.

spectrograph analysis　An analysis to determine elements present in an unknown. It may be quantitative or qualitative.

spidering　As applied to soldering, refers to a condition wherein the plastic basic material of the printed circuit board substrate is softened as it passes over the solder wave, with a resultant pick up of fine particles of solder onto the tacky surface of the plastic. This is generally the result of inadequate curing of the plastic material comprising the substrate. Particular difficulties result in printed circuit boards with closely spaced conductors and/or with high voltage present. Also known as webbing.

spot　On a printed circuit, it is the conductive area to which components or separate circuits are attached. A plated-through hole land is the conductor surrounding the hole through the conductive pattern and the base material. Also called boss, pad, terminal point, tab, land, or donut.

spray fluxing　Method of applying flux during wave-solder operation in which the liquid flux is sprayed onto the underside of the PCB assembly.

spring-loaded test probes　Head styles of probes include serrated, center point crown, chisel, crown, cup spear, and concave.

squeegee　A rubber or metal blade used in screen and stencil printing to wipe across the screen/stencil to force the

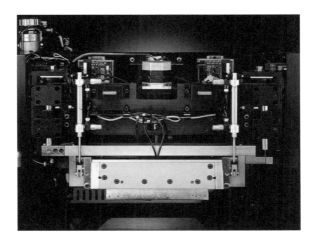

Figure 11 Head assembly containing squeegee on automatic stencil printing machine.

solder paste through the screen mesh or stencil apertures onto the footprint of the PCB.

squeegee downstop The travel in the Z-axis of the squeegee to the stencil.

squeegee pressure The pressure exerted on the stencil during the print cycle, typically in the range of one to six pounds per squeegee inch. Also known as down force.

squeegee speed The rate at which the squeegee blade moves across the stencil during the print cycle. Typical speeds range between 0.4 to 6.0 in./sec. Also known as print speed.

squeegee stroke The length in inches of squeegee blade travel over the stencil.

stabilization period Period of time after preheat and before the reflow spike where the internal temperature differences between components are allowed to equalize. Also known as drying time, preflow, and soak.

staggered pattern A solder paste stencil configuration used to control the volume of solder paste applied to fine pitch component land patterns to avoid excessive solder joints. The pattern would alternate from pad to pad the placement of the solder from the heel region to toe region.

stainless steel screen A stainless steel mesh screen stretched across a frame and used to support a circuit pattern defined by an emulsion bonded to the screen.

stair-step print A print that retains the pattern of the screen mesh at the line edges. This is a result of inadequate dynamic printing pressure exerted on the paste or insufficient emulsion thickness coating the screen.

standard pitch The center-to-center lead distance of surface mount packages that are 0.035 inch or more.

static dissipative materials ESD protective materials having surface resistivities greater than 10^5 but not greater than 10^9 ohms per square centimeter.

statistical process control (SPC) The use of statistical techniques to analyze a process or its output to determine any variation from a benchmark and to take appropriate action to restore statistical control, if required.

stencil A thick sheet material with a circuit pattern cut into the material. A metal mask is a stencil. The most common material is stainless steel.

stencil printer Machine that dispenses solder paste or other material by means of a squeegee forcing material through the apertures of a stencil. Equipment ranges from manual to semiautomatic to automatic. (See Figure 13 on page 100.)

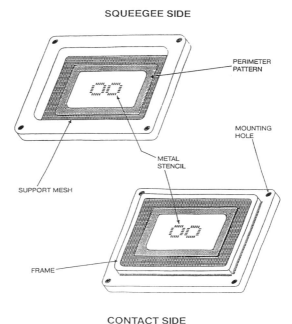

Figure 12 Stencil. (Courtesy ITM, Inc.)

stencil wiper Peripheral device on some automatic stencil printers that automatically wipes the underside of the stencil between printing cycles. (See Figure 14.)

step-and-repeat A process wherein the conductor or resistor patterns are repeated many times in evenly spaced rows onto a single film or substrate.

step-down stencil A stencil of varying thickness to accommodate coarse and fine-pitch deposits. Also known as multi-level stencil.

stonehenge effect A soldering defect in which a chip component has been pulled into a vertical or near vertical position with only one termination soldered to the PCB. It is typically caused by force imbalances during the re-

Figure 13 Automatic stencil printing machine. (Courtesy MPM Corporation)

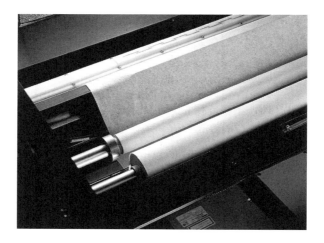

Figure 14 Stencil wiper as incorporated on automatic stencil printing machine. (Courtesy MPM Corporation)

flow soldering process. Also referred to as drawbridging, tombstoning, and the Manhattan effect.

storage life The period of time during which a liquid resin or adhesive can be stored and remain suitable for use. Also called shelf life.

stress relief A method used to minimize stresses to the soldered termination or component usually in the form of a bend or service loop in a component lead and a solid or stranded wire to provide relief from stresses between terminations caused by movement, thermal expansion, or other external factor.

stripping Process of removal of resist.

subpanel A PCB panel containing multiple board images and used as the basic unit for assembly processing, as opposed to the full panel prepared for bare board fabrication. Hence, two or more subpanels result from the full panel.

substrate The base material that forms the support structure of an IC or PCB.

subtractive process A process for obtaining conductive patterns by the selective removal of unwanted portions of a conductive foil.

surface conditioners Specially formulated liquid cleaners to restore the solderability of the most commonly used metals and alloys in the soldering process.

surface insulation resistance The electrical resistance of the insulating material, determined under specified environmental and electrical conditions, between a pair of contacts, conductors, or grounding devices in various combinations, measured in ohms. Also referred to as SIR.

surface mount assembly (SMA) Electronics assembly in which non-through-hole components are mounted directly to the substrate surface.

102 surface mount component (SMC)

surface mount component (SMC) A component designed to be mounted and soldered to pads on the surface of a PCB rather than inserted into through-holes in a PCB.

surface mount device Electronic components, either active (transistors, integrated circuits, diodes, etc.) or passive (capacitors, resistors, coils, etc.) that do not have wire leads or pins. The terminal leads are part of the component body, thus allowing direct mounting on the surface of printed circuit boards. Leaded devices are mounted by their leads through holes drilled in the boards. In both cases, the components are held in place on the boards, both mechanically and electrically, by solder. Surface mounted components usually are smaller than an equivalent through-hole leaded device and, in some cases, less expensive. Equipment and design engineers can save valuable wiring board area by mounting surface-mounted leadless components on the underside of a board, and conventional leaded components on the top side of boards. Abbreviated SMD.

surface mounting Electrical connection of components to the surface of a conductive pattern without using through-holes.

surface tension A property of liquids, due to molecular forces existing in the surface film of all liquids, which tends to contract the volume into a form with the least surface area. That is, the molecules on the surface of a liquid are not acted upon by the same forces as those molecules in the interior of the liquid.

surfactant A chemical added to water to lower the surface tension and to improve wetting. It is a short form for "surface active agent."

suspension A mixture of liquid or solid in a liquid. It is not considered a true solution because discrete particles or droplets are visible and the mixture is not clear.

synthetic activated flux A high activity organic flux whose post soldering residues are soluble in commonly used halogenated solvents. Abbreviated SA.

syringe A tool that also packages a paste or epoxy in a container for manual, hydraulic, or pneumatic deposition through a needle opening.

syringe dispensing Material deposition by applying pressure (pneumatic or hydraulic) for a certain period of time to expel the appropriate amount of material through the needle to the target location.

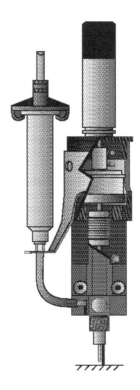

Figure 15 Dispensing pump from automatic material dispensing machine used to dispense adhesives and solder paste. (Courtesy Camelot Systems, Inc.)

tab   On a printed circuit, it is the conductive area to which components or separate circuits are attached. A plated-through hole land is the conductor surrounding the hole through the conductive pattern and the base material. Also called boss, pad, terminal point, land, spot, or donut.

tackiness   The ability of the solder paste to hold the surface mount components in place after the component placement operation but before the reflow soldering process.

tail   A flexible circuit attached to a rigid printed circuit board.

tape automated bonding   A technology for gang bonding a lead frame directly from the silicon die to the PCB. Abbreviated TAB.

tape-and-reel   Component packaging method of housing each component in its own tape cavity in a long continuous strip. The tape cavities are covered with a plastic sheet so the tape can be wound around a reel for component presentation or "feeding" to the automated component placement machine.

taped components   Components attached to a continuous tape for use in automatic assembly. Also referred to as tape-and-reel.

temperature   A measure of the energy held by the molecules of a material.

temperature cycling An environmental test where a component or assembly is subjected to several alternating temperature cycles from a low temperature to a high temperature to a low temperature, etc., over a period of time.

temperature profile A graphic depiction of the temperature in which a selected point traverses as it passes through the reflow process.

temperature rating The maximum temperature at which the insulating material may be used in continuous operation without loss of its basic properties.

temperature, preheat The final temperature a selected point achieves in the preheat process.

temperature, ambient Dry bulb temperature of area under consideration.

temperature, flux activation A point of elevated temperature that causes the flux to become active in removing oxides on the metals to be joined.

temperature, liquidus The temperature above which an alloy is completely liquid.

temperature, maximum reflow The maximum temperatures that any point on a board will reach at reflow conditions.

temperature, reflow Often listed in reference to a band of temperatures where solder reflow takes place. For tin-lead it is typically 205°–220°C.

temperature, saturation The boiling point of a fluid.

temperature, solidus The point above which solder first melts.

tensile strength A material characteristic describing its resistance to fracture when the material is stretched.

tenting A PCB fabrication technique whereby the plated via holes and/or surrounding conductive patterns are covered with a dry film solder resist.

terminal area A portion of a conductive pattern usually, but not exclusively, used for the connection and/or attachment of components.

terminal point On a printed circuit, it is the conductive area to which components or separate circuits are attached. A plated-through hole land is the conductor surrounding the hole through the conductive pattern and the base material. Also called boss, pad, land, tab, spot, or donut.

termination The part of a component that makes electrical contact with the PCB.

test board A printed board suitable for determining acceptability of the board or of a batch of boards produced with the same process so as to be representative of the production board.

test coupon A portion of functioning circuitry used exclusively for testing and positioned for access.

test coupon A sample or test pattern usually made as an integral part of the printed board, on which electrical, environmental, and microsectioning tests may be made to evaluate board design or process control without destroying the basic board.

test pattern A circuit or group of substrate elements processed on or within a substrate to act as a test site or sites for element evaluation or monitoring of fabrication processes.

test rate A component placement term similar to the placement rate, except it measures the machine speed as it moves from the component pickup location to the board site location and back again without actually picking and placing components (a dry run). Also referred to as the cycle rate.

thermal bridge   Metal cover over components mounted on a PCB with a thermal plane. Specifically applied to aid heat dissipation.

thermal coefficient of expansion   The rate of expansion of a material, measured in ppm/°C when the material's temperature is increased. Abbreviated TCE. The TCE of a substrate must closely match the TCE of a solidly mounted SMC to prevent mechanical stress on the solder joints.

thermal conductivity   That property of a material or assembly that describes its ability to conduct heat. Metals, in general, are better thermal conductors than non-metals. Silver and copper are the best conductors of heat. In general, thermal conductivity of a material parallels its electrical conductivity.

thermal design   PCB design philosophy that ensures adequate heat dissipation. Also known as thermal management.

thermal insulation   The inverse of thermal conductivity. The ability of a material to block or resist the flow of heat.

thermal management   Consideration and use of the various methods for dissipating heat produced by operating electronic equipment so that the operating temperature specification of the components is not exceeded.

thermal mismatch   Differences of thermal coefficients of expansion of materials that are bonded together.

thermal plane   A heatsink bonded to the surface of a PCB before component insertion or laminated within a PCB to aid heat dissipation, specifically in closely packed PCB assemblies.

thermal relief Crosshatching to minimize blistering or warping during soldering operations.

thermal shock Condition whereby devices are subjected alternately to extreme heat and cold. Used to screen out processing defects.

thermal via A plated via under a component that is used to conduct heat away from the component.

thermocouple A device for measuring temperature where two conductors of dissimilar metals are joined at the point of heat application, yielding a voltage difference proportional to the temperature.

thermocouple Made of two dissimilar metals which, when heated, generate a small DC voltage. Typically used in measuring temperatures.

thermode A heating element utilizing electrical current to generate heat for contact reflow soldering.

thermode temperature gradient The difference in temperature between the highest and lowest temperature as measured from one end of a thermode to the other at a given instant in time once steady state temperatures are attained.

thermode temperature variation The difference between the highest and lowest thermode temperature as measured at one point when controlled to one temperature over a period of time.

thermode-thermode variation The maximum difference in thermode temperature between the same point on each thermode of a multiple thermode tool.

thermoplastic Plastics that can be repeatedly made to flow under application of heat to fill a mold, coat non-plastic materials, extrude shapes, etc.

thermoset Materials that chemically change upon curing as a result of heat or catalytic action. The process is irreversible.

thick-film technology Technology whereby electrical networks or elements are formed by applying, and subsequently firing, a liquid, solid, or paste coating through a screen or mask in a selective pattern onto a supporting material (substrate) and fired.

thixotropic The characteristic of a liquid or gel that is viscous when static, yet fluid when physically worked.

time above liquidus The time that a selected solder joint is above the liquidus temperature in the reflow process.

tinning The coating of a terminal, lead, or conductive pattern with tin or solder alloy to improve or maintain solderability or to aid in the soldering operation.

tinning oils Liquid compounds formulated for use as the oil in oil intermix wave soldering equipment and as pot coverings on still solder pots. They provide a barrier between the atmosphere and molten solder, thereby reducing the solder oxidation (drossing), and also reduce the surface tension of the molten solder, thereby enhancing the wetting characteristics.

tombstoning A soldering defect in which a chip component has been pulled into a vertical or near vertical position with only one termination soldered to the PCB. It is typically caused by force imbalances during the reflow soldering process. Also referred to as drawbridging, the Manhattan effect, and the stonehenge effect.

tooling feature A specified physical feature on a printed circuit board or a panel, such as a marking hole, cut-out, notch, slot, or edge, used exclusively to position the board or panel or to mount components accurately.

tooling holes Holes placed on a PCB or panel and used to aid in accurately positioning the substrate on a machine center.

toxicity Relates to all forms of human exposure to substances that can cause distress: inhalation, skin contact, and ingestion.

trace A thin conductive area on a PCB or internal layer, usually composed of paths and lands to which component leads are connected. Also referred to as a conductor.

transistor A current amplifying semiconductor.

translational error The misalignment of the component centroid resulting from inaccuracies of an X-Y positioning system including offset, scaling, and axis orthogonality errors.

transmissivity Percentage of incident infrared energy that is transmitted through a defined thickness of material.

traveler Set of instructions that lists all the steps the PCB will go through during manufacturing and the requirements of each step. It is consulted and read before any work is performed on the assembly and signed off when the work has been completed (along with the actual quantity of assemblies written in).

true position The exact location of a feature or hole established by basic dimension.

true position tolerance The total diameter of permissible movement around the true position as shown in the master drawing.

tube feeder A component packaging method of inserting components back to back in an anti-static or conductive plastic tube or stick. They may use vibration, gravity, or spring action to facilitate the component indexing to the pick up tool.

twang test A method of determining the tackiness of solder paste over time, i.e., its ability to hold components in place. The assembly is deflected 1" then released (twang!). The favorable result is to have all components still remaining in their specific locations within the solder paste.

twist The deformation of a rectangular sheet resulting in one of the corners not being in the plane containing the other three corners.

Type I Assembly An exclusive SMT PCB assembly with components mounted on one or both sides of the substrate.

Type II Assembly A mixed technology PCB assembly with SMT components mounted on one or both sides of the substrate and through hole components mounted to the primary or component side.

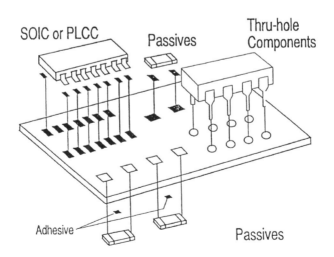

Figure 16 Type II PCB assembly. (Courtesy ITM, Inc.)

Type III Assembly A mixed technology PCB assembly with passive SMT components mounted on the secondary side of the substrate and through-hole components mounted to the primary or component side. Typically this type of assembly is wave soldered in a single pass.

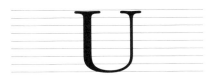

ultra fine pitch The center-to-center lead distance of surface mount packages that are 0.010" or less. Abbreviated UFP.

ultrasonic cleaning equipment Devices used for ultrasonic immersion that include a transducer that converts electrical energy into mechanical energy, an ultrasonic generator, and a tank to contain the cleaning liquid. Systems can be batch or conveyorized.

ultrasonic immersion Cleaning technique utilizing cavitation (rapid formation of tiny bubbles in a cleaning liquid). Cavitation is created by ultrasonic, high intensity sound waves. The agitation of imploding bubbles scrubs the immersed object.

ultrasonics A technology for cleaning residues utilizing cavitation (microagitation via sound waves) by means of a high-frequency generator and a transducer.

undercut The reduction (narrowing) of conductor cross-section on a printed circuit caused by etchant removal of conductive metal under the edge of the resist.

underfill Encapsulant material that is deposited under a device (e.g., a flip-chip on a plastic substrate) to help compensate for a mismatch in TCE between the device and the substrate.

Underwriters Laboratories, Inc. An independent testing organization that evaluates the performance and capability of electrical wiring and equipment to ensure that they meet standards for safety.

unit under test (UUT) A term applied to any circuit board that is being tested by ATE.

universal grid access fixture A fixture used for testing bare PCB substrates' inner layers and backplanes. Facilitates testing, fabrication, and assembly.

upload The ability of a system to accept program data from a host computer.

UV curing Polymerizing, hardening, or cross linking a low molecular weight resinous material in a wet coating or ink, using ultraviolet light as an energy source. UV techniques allow inks to be 100 percent convertible. UV is fast, requiring relatively small equipment space and low energy requirements.

vacuum pickup A handling system featuring a small vacuum cup at one end for securing parts in a placement movement.

vacuum test fixture Used to test both bare and loaded PCB assemblies on a dedicated test system.

vapor degreaser Cleaning system in which an object is suspended in a chamber while heated solvent vapors condense on its surfaces. Repeated flushing is achieved with fresh solvent.

vapor phase (1) A general term referring to condensation heating where the object to be heated is submerged into a hot, saturated vapor. The object, being cooler than the vapor, causes the vapor to condense on the object, thus transferring its latent heat of vaporization to the part. This is an equilibrium method of heating as the part temperature becomes asymptotic to the vapor temperature over a short period of time. Also known as condensation inert heating. (2) The state of a compound when it is in the form of a vapor.

vapor phase soldering Vapor phase (condensation) soldering uses the high-temperature vapor of a boiling perfluorinated hydrocarbon as the heat transfer medium. Also known as condensation inert soldering.

vapor pressure That pressure exerted by a vapor in equilibrium with a solution, or the material from which the vapor emanated. Vapor pressure is dependent upon temperature, i.e., the higher the temperature, the higher the vapor temperature. It is indicative of the volatility of a material, i.e., the higher the vapor pressure of a material at a given temperature, the higher will be its evaporation rate. In terms of fluxes and cleaning solvents used in the soldering process, these materials are generally mixtures of materials. The components with the highest vapor pressure will evaporate faster than the other components, eventually resulting in an imbalance of the flux or solvent.

vapor recovery The process of retrieving working fluid vapors and aerosols from scavenged air and turning it into reusable working fluid.

vapor, saturated The condition of a vapor when it is in equilibrium with its boiling fluid.

vaporization, heat of The amount of heat required to change a given amount of saturated fluid into saturated vapor.

vehicle Thick film term that refers to the organic system in the paste.

vendor assessment Buying procedure in which the manufacturer of parts assesses the product, therefore the purchaser does not have to do so prior to use.

very fine pitch (VFP) The center-to-center lead distance of surface mount packages that are between 0.012 inch and 0.020 inch.

VHSIC Very High Speed Integrated Circuit.

via 1) A vertical conductor or conductive path following the interconnection between multi-layer circuit layers. 2) An interlayer connection in a multi-layer board.

via hole A plated-through hole used as a through connection, but in which there is no intention to insert a component lead or other reinforcing material.

via, blind A hole that is connected to either the primary or secondary side of a multi-layer assembly, but not to both.

via, buried A hole that is connected to neither the primary nor the secondary side of a multi-layer assembly, i.e., it connects only between inner layers.

viscosity The measure of resistance of a fluid to flow (either through a specific orifice or in a rotational viscometer). The absolute unit of viscosity measurement is poise, more commonly centipoise. Viscosity varies inversely with temperature.

visible light (band) Electromagnetic radiation occurring between the wavelengths of 0.39 and 0.78 microns. Produces light and heat.

vision system One or more cameras used to center SMT components for accurate placement onto the PCB.

VLSI Very Large Scale Integration. VLSI devices are ICs that contain 1000 or more gate equivalents.

voids Cavities inside the solder joint formed by gases that are released during reflow or by flux residues that fail to escape from the solder before it solidifies.

volatile Used to describe materials that have a relatively high evaporation rate or a tendency to evaporate.

warp and woof Threads in a woven screen that cross each other at right angles.

warpage The distortion of a substrate from a flat plane.

water based A description of a liquid system where the primary solvent is water.

water displacement Characteristic of certain materials, such as lacquers and protective coatings, that replace water.

water extract resistivity Value in ohm-centimeters, principally for liquid rosin fluxes, obtained by carrying out a standard test that measures the amount of ionizable material present. The higher the value, the higher the resistivity, hence the less ionizable material present.

water softener A system used to remove deposit-forming ions, by replacing them with non-scaling sodium ions. This low cost process is used in residential as well as commercial applications.

water soluble A description of a liquid system, where the prime solvent is not necessarily water. However, the system is soluble in water, i.e., can be dissolved in or by water.

water soluble organic flux A nonspecific term describing the composition of soldering of flux soluble in water and based on organic rather than inorganic chemicals: organic acids, amine hydrohalides, polyglycols, polyhydric alcohols, etc.

wattage A unit of electrical power determined mathematically by multiplying the electrical current by the voltage. Soldering irons are generally classified by wattage, which is indicative of the rate at which they will solder.

wave soldering The technique of soldering components to a substrate by passing the PCB assembly over a wave of molten solder in a soldering pot. The wave is maintained above the level of the pot by solder being pumped through a manifold in the bottom of the pot.

weave exposure Unbroken fibers of woven glass cloth that are not completely covered by resin on the base material surface.

webbing As applied to soldering, refers to a condition wherein the plastic basic material of the printed circuit board substrate is softened as it passes over the solder wave, with a resultant pick up of fine particles of solder onto the tacky surface of the plastic. This is generally the result of inadequate curing of the plastic material comprising the substrate. Particular difficulties result in printed circuit boards with closely spaced conductors and/or with high voltage present. Also known as spidering.

wetting A physical phenomenon of liquids, usually in contact with solids, wherein the surface tension of the liquid has been reduced so that the liquid flows and makes intimate contact in a very thin layer over the entire substrate surface. Regarding wetting of a metal surface by a solder, flux reduces the surface tension of the metal surface and the solder, resulting in the droplets of solder collapsing into a very thin film, spreading, and making intimate contact over the entire surface.

wetting agent A chemical material added to a liquid solution to reduce surface tension. The effect of this reduction of surface tension is to increase the power of the liquid mixture or solution to wet an object on which it is placed.

wetting balance Method of assessing solderability of metals.

whiskers A growth in the presence of condensed moisture and electrical bias resulting in metallic filaments between conductors. Also called dendritic growth.

wicking The flow of solder along conductors under insulation through via holes.

working fluid Usually a fully perfluorinated hydrocarbon that is extremely stable and is clear, odorless, and chemically inert.

worknest The stage that holds the substrate throughout the print cycle.

x-ray imaging Optical test procedure using the ability of x-rays to pass through certain substances more easily than others.

yield, test The ratio (in percent) of boards that pass a level of testing to the total tested.

yield, production The ratio (in percent) of usable assemblies or components at the end of a manufacturing process to the number of components initially submitted for processing.

yield loss The ratio (in percent) of product that fails to meet a level of testing.

Z Symbol for impedance, designated in ohms.

Z-axis adhesive A material filled with a low concentration of large conductive particles designed to conduct electricity in the Z-axis but not the X- or Y-axis. Also called an anisotropic adhesive.

Z-stroke The movement of the placement head of a component placement machine in the vertical plane (up and down) for component orientation and placement onto the PCB.

zero-ohm resistor An automatically insertable jumper wire used to connect between two points of a PCB. It resembles a resistor and is often used on single-layer boards as a means of simplifying conductor routing.

Abbreviations and Symbols

°	angular measure; degree
°C	degree Celsius/centigrade
°F	degree Fahrenheit
BGA	micro Ball Grid Array
3DPM	3 Dimensional Packaging Module
A	ampere
AAGR	average annual growth rate
ABS	Acrylonitrile-Butadiene-Styrene (plastic)
ac	alternating current
AC	alternating current
ACI	automatic component insertion
AEC	architecture, engineering, and construction
AFNOR	Association Française de Normalization
AGV	automatic guided vehicle
AI	artificial intelligence
AIS	adhesive interconnect system
ANOVA	analysis of variance
ANSI	American National Standards Institute
AOI	automatic optical inspection
AOQ	average outgoing quality
APT	automatically programmed tools
AQL	acceptable quality level
ARPA	Advanced Research Projects Agency
ASCII	American Standard Code for Information Interchange
ASIC	application specific integrated circuit
ASM	American Society for Metals

Abbreviations

ASME	American Society of Mechanical Engineers
ASQC	American Society for Quality Control
ASTM	American Society for Testing Materials
at. %	atomic percent
ATE	automatic test equipment
ATG	automatic test generation
atm	atmospheres
AWG	American Wire Gauge
BGA	ball grid array; bumped grid array
BITE	built-in test equipment
BOM	bill of material
BS	boundary scan
BTAB	bumped tape-automated bonding
C4	controlled collapse chip connection
CAD	computer aided design
CAE	computer aided engineering
CAFM	computer aided facilities management
CAGE	commercial and government entity
CALS	computer aided acquisition and logistic support
CAM	computer aided manufacturing
CAPP	computer aided process planning
CAR	computer aided repair
CASE	computer aided software engineering
CAT	computer aided testing
CBGA	ceramic ball grid array
CCA	circuit card assembly
CCAPS	circuit card assembly and processing system
CCD	closed captioned device
CFC	chlorofluorocarbon
cfh	cubic feet per hour
CFM	continuous flow manufacturing
cfm	cubic feet per minute

CIC	condensation inert curing
CIH	condensation inert heating
CIM	computer integrated manufacturing
CIS	condensation inert soldering
CLCC	ceramic leaded chip carrier
CMA	circuit mil area
CMOS	complementary metal oxide semiconductor
CNC	computer numerical control
COB	chip-on-board
COD	consumed oxygen demand
Cp	capability performance
CPBGA	cavity plastic ball grid array
CPI	continuous process improvement
CPL	capability performance—lower
CPU	capability performance—upper
CPU	central processing unit (computer)
CRT	cathode ray tube
CSA	Canadian Standards Association
CSP	chip scale package
CSP	chip size package
CTE	coefficient of thermal expansion
D	diameter
dB	decibel
dc	direct current
DC	direct current
DFM	design for manufacturability
DFMA	design for manufacturability and assembly
DFT	design for test
diam	diameter
DIL	dual inline
DIN	Deutsches Institut für Normung
DIP	dual inline package

Abbreviations

DNC	direct numerical control
DOD	Department of Defense
DOS	disc operating system
dpm	defects per million
DRAM	dynamic random access memory
DRC	design rule checking
DUT	device under test
DVM	digital voltmeter
ECAD	electronic computer aided design
ECCB	Electronic Component Certification Board
ECL	emitter coupled logic
ECN	engineering change notice
ECO	engineering change order
EDA	electronic design automation
EDF	electronic design interchange format
EIA	Electronic Industries Association
EIAJ	Electronics Industry Association in Japan
EIS	engineering information system
EMC	electromagnetic compatibility
EMF	electromotive force
EMI	electromagnetic interference
EMP	electromagnetic pulse
EMPF	Electronics Manufacturing Productivity Facility
EPA	Environmental Protection Agency
ESD	electrostatic discharge
ESS	environmental stress screening
FAA	Federal Aviation Administration
FAC	forced air convection
FAR	failure analysis report
FCC	Federal Communications Commission
FCC	flat conductor cable
FCT	functional test card

Abbreviations

FEA	finite element analysis
FEM	finite element modeling
FET	field effect transistor
FLT	full liquidus temperature
FOP	fineness of print (stencil)
FPBGA	fine pitch ball grid array
FPT	fine pitch technology
FTIR	Fourier transform infrared spectroscopy
GaAs	gallium arsenide
HAL	hot air leveling
HASL	hot air solder leveling
I/O	input/output
IC	integrated circuit
ICAM	integrated computer aided manufacturing
ICEA	Insulated Cable Engineers Association
ICT	in-circuit testing
ID	inner diameter
IEC	International Electrotechnical Commission
IECQ	International Electronic Component Qualification System
IEEE	Institute of Electrical and Electronic Engineers
IEPS	International Electronic Packaging Society
ILB	inner lead bonding
IMC	intermetallic compound
IPC	Institute for Interconnecting and Packaging Electronic Circuits
ipm	inches per minute
IR	infrared
ISA	Instrument Society of America
ISHM	International Society for Hybrid Microelectronics
ISO	International Standards Organization
JEDEC	Joint Electronic Device Engineering Council

JIT	just in time
JTAG	Joint Test Action Group
KGB	known good board
KGD	known good die
LAN	local area network
LCCC	leadless ceramic chip carrier
LCL	lower control limit
LEO	lands exposed only
LGA	land grid array
LID	leadless inverted device
LIF	low insertion force
LN	Deutsche Luft und Raumfahrt Norm
LSI	large scale integration
MAP	manufacturing automation protocol
MCAD	mechanical computer aided design
MCAE	mechanical computer aided engineering
MCB	molded circuit board
MCM	multi-chip module
MCM-C	multi-chip module ceramic
MCM-D	multi-chip module deposited
MCM-L	multi-chip module laminate
MDA	manufacturing defects analyzer
MELF	metal electrode leadless face
MIL	military
miniBGA	mini ball grid array
MIR	moisture insulation resistance
MLB	multi-layer board
MLPWB	multilayer printed wiring board
MOS	metal oxide semiconductor
MRP	manufacturing resource planning
MRP	material requirement planning
MSI	medium scale integration

Abbreviations 131

MTBF	mean time between failure
MTTF	mean time to failure
MTTR	mean time to repair
MVC	most vulnerable component
NADCAP	National Aerospace and Defense Contractors Accreditation Procedures
NASA	National Aeronautics and Space Administration
NBS	National Bureau of Standards
NC	numerical control
NDT	non-destructive testing
NEMA	National Electrical Manufacturers Association
NFPA	National Fire Protection Association
NIST	National Institute for Science and Technology
NMR	normal mode rejection
OA	organic acid
OD	outer diameter
OEM	original equipment manufacturer
OLB	outer lead bonding
OSHA	Occupational Safety and Health Administration
P/I	packaging and interconnection
PAC	pad array carrier
PBGA	plastic ball grid array
PCB	printed circuit board
PGA	pin grid array
PLCC	plastic leaded chip carrier
PNP	pick and place
ppb	parts per billion
ppm	parts per million
ppt	parts per trillion
psi	pounds per square inch
psia	pounds per square inch absolute

132 Abbreviations

psig	gauge pressure in pounds per square inch (pressure relative to ambient pressure)
PTH	plated through hole
PWB	printed wiring board
QAS	quality assessment schedule
QFP	quad flat pack
QPL	qualified products list
RA	rosin fully activated
RAM	random access memory
RMA	rosin mildly activated
ROM	read only memory
ROT	reflow of through-hole
rpm	revolutions per minute
RSA	rosin super activated
SA	synthetic activated
SA0	stuck at 0
SA1	stuck at 1
scfh	standard cubic feet per hour
SCRS	single center reflow soldering
SEM	scanning electron microscope/microscopy
sfm	surface feet per minute
SI	Système International d'Unités
SIP	single in-line package
SIR	surface insulation resistance
SLICC	slightly larger than integrated circuit carrier
SMA	surface mount assembly
SMART	Surface Mount and Related Technologies (British SMT Organization & Suppliers)
SMC	surface mount component
SMD	surface mount device
SME	Society of Manufacturing Engineers
SMOBC	solder mask over bare copper

Abbreviations 133

SMT	surface mount technology
SMTA	Surface Mount Technology Association
SO	small outline
SOIC	small outline integrated circuit
SOJ	small outline j-lead (component)
SOL	small outline large (body)
SOP	small outline package
SOT	small outline transistor
SOW	small outline wide (body)
SPC	statistical process control
SPRS	single pass reflow soldering
SQC	statistical quality control
SSOP	slim small outline package
T	temperature
TAB	tape automated bonded
TBGA	tape ball grid array
TCE	thermal coefficient of expansion
TCP	tape carrier package
TQFP	thin quad flat pack
TQM	total quality management
TSOP	thin small outline package
TSSOP	thin shrink small outline package
UCL	upper control limit
UFPT	ultra fine pitch technology
UL	Underwriters Laboratories
UUT	unit under test
V	volt
VDE	Verband Deutscher Elektrotechniker
VG	Verteidigungsgerte Norm
VLSI	very large scale integration
VOC	volatile organic compound
vol	volume

VPS	vapor phase soldering
VSPA	very small peripheral array
W	watt
WI	wettability index

Related Newnes Titles

Modern Dictionary of Electronics
Sixth Edition
Rudolf Graf
This is a classic, comprehensive reference book for engineers, technicians, students, and hobbyists. It includes practical terminology for consumer electronics, optics, microelectronics, communications, medical electronics and packaging and production.
1996 • 1,152pp • Paperback • 0 7506 9870 5 • $49.95

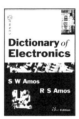

Dictionary of Electronics
Third Edition
S.W. Amos and Roger Amos
The third edition of this dictionary includes over 4,000 terms covering basic technology and applications. More than 400 of these are new for the third edition. Acronyms and abbreviations are also included.
1996 • 384pp • Paperback • 0 7506 2405 1 • $29.95

Electronics Terminology: A Concise Dictionary
Informatik Rezurch
This book consists of two sections: acronyms and glossary. It features electronics, packaging and production, computer and semiconductor terms and definitions which are current and straight to the point.
1996 • 96pp • Paperback • 0 7506 9751 2 • $12.95

These books are available from all better book stores or in case of difficulty call: 1-800-366-2665 in the U.S. or +44 1865 310366 in Europe.

Detailed information on these titles may be found in the Newnes catalog (Item #645). To request a copy, call 1-800-366-2665. You can also visit our web site at: http://www.bh.com/newnes.

E-Mail Mailing List
An e-mail mailing list giving information on latest releases, special promotions/offers and other news relating to Newnes titles is available. To subscribe, send an e-mail message to majordomo@world.std.com. Include in message body (not in subject line) subscribe Newnes.